THE SEAWEED REVOLUTION

The Seaweed Revolution

How Seaweed Has Shaped Our Past and Can Save Our Future

Vincent Doumeizel

Translated by
Charlotte Coombe

Illustrations by
Neige Doumeizel

HERO, AN IMPRINT OF LEGEND TIMES GROUP LTD
51 Gower Street
London WC1E 6HJ
United Kingdom
www.hero-press.com

First published in French as *La Révolution des algues* by Éditions des
Équateurs/Humensis in 2022
This translation first published by Hero in 2023

© Éditions des Equateurs/Humensis, 2022
Translation © Charlotte Coombe, 2023

The right of the author and translator to be identified as the author
and translator of this work has been asserted in accordance with the
Copyright, Designs and Patents Act 1988. British Library Cataloguing
in Publication Data available.

This translation was funded by the Lloyd's Register Foundation

Printed and bound in Great Britain by TJ Books Ltd, Padstow

ISBN (HARDBACK): 978-1-91564-385-8
ISBN (SPECIAL EDITION): 978-1-91564-346-9

The Seaweed Revolution

Rope-grown seaweed.

Introduction

Seaweed is often misunderstood and seen as a form of pollution, when it is only a symptom of it. In fact, seaweed offers an endless source of innovation and concrete solutions that could help us address some of the major challenges facing our generation.

Our society has spurned seaweed, undoubtedly the world's greatest untapped resource, but the climate emergency and global population growth are now pushing us to reconsider this overlooked treasure.

If we learn how to grow it sustainably, seaweed could feed people, replace plastic, decarbonize the economy, cool the atmosphere, clean up the oceans, rebuild marine ecosystems and reduce social injustice by providing jobs and income to coastal populations where fishing resources are disappearing.

An essential pillar of life on earth, seaweed reproduces quickly and can grow dozens of metres in a few days without needing food, fresh water or pesticides.

It's time to dive into this green, brown and red ocean filled with unexplored resources. If we want to rebuild ecosystems instead of destroying them, seaweed is an excellent place to start.

Our ancestors, algae...

The first form of life on this planet was algae. Born out of a process of photosynthesis that has existed for 3.5 billion years, single-celled blue bacteria evolved in symbiosis with other cells to produce microalgae. Essential to life on earth, they are our most distant ancestors. Made up of single cells, these microalgae evolved and became more complex, forming multicellular organisms. And so, over a billion years ago, macroalgae appeared.

These plants have fed humans for tens of thousands of years and contributed to making us *Homo sapiens*. They are a fundamental element of our existence on earth. More than half of the oxygen present in the atmosphere comes from the oceans, and therefore from algae. Without them, there would be no shells or marine fauna. The ocean would be little more than a desert, incapable of absorbing carbon or producing oxygen.

There are about 12,000 species of macroalgae in the oceans. Green and red, these life forms have adapted over the course of their evolution and have taken very diverse paths. The only things they have in common are the way they grow (photosynthesis) and the ecosystem in which they evolve (water).

Around a billion years ago, some primitive red algae transformed to create the brown algae family, which includes the kelp often found washed up on beaches.

As for green algae, some of them were spewed out onto the coastlines 500 million years ago and have transformed into all the terrestrial vegetation we know today. Oaks, strawberry plants, roses and baobabs are all descended from green algae.

Even today, green algae are still genetically closer to tomato plants or fir trees than they are to red algae. Red algae and green algae share no common ancestors in the plant world. They are more genetically different from one another than a mushroom and a bear!

For centuries, we have neglected these marine plants, both from a food point of view as well as an industrial or environmental one. European culture certainly does not value seaweed very highly. Virgil stated in the *Aeneid*: '*Nihil vilior alga!*' ('There is nothing more worthless than seaweed!'), while Aristotle placed them last in the 'great chain of being'. One ancient proverb even declares that 'the sea hates seaweed so much that she throws it back onto the beach'. In all the territories they invaded, Europeans erased the traditions related to these plants. The Native American peoples, the Māori and the Australian Aborigines all ate seaweed before they were colonized. Only Japan has managed to maintain its appetite for these foods. Maybe it can be seen as an aesthetic and ecological prejudice: initially considered an unsightly waste on beaches, the proliferation of seaweed has also recently been described as green and red 'invasions'.

Yet, anyone who dives into one of the giant underwater forests that can still be found off the coast of Chile, California or Australia will be amazed by the splendour of these plants. Their gigantic lianas sway in the water, reaching up to dozens of metres in height, criss-crossed by rainbow-coloured fish and beams of light, in majestic silence.

What is seaweed?

A seaweed, or alga,[1] has no differentiated cells: unlike a plant, it does not have roots, flowers or sap. It consists of a holdfast, a stipe and fronds (or blades). The holdfast allows it to cling to a substrate but, unlike roots, does not absorb any nutrients. The stipe is the equivalent of a plant's stem and allows the frond to move up towards the light. The frond is the part that is visible in the water and often the part that is eaten. It plays a role equivalent to that of a leaf. Seaweed may lose its fronds and regrow others.

Contrary to its name, it isn't actually a weed, although some green seaweeds are the ancestors of land-based weeds and grasses. In Asia and increasingly around the world, seaweed is referred to as 'sea vegetables' or 'sea greens', descriptions which better reflect its nature, flavours and benefits.

Perhaps one day, when we have grasped how crucial it is to protect these gigantic ecosystems, we will call them 'sea forests'. Preserving them is just as necessary to life on earth as saving the forests of the Amazon or elsewhere.

In this book, we will be focusing on macroalgae – their advantages, benefits and uses for our food, our health and our planet, and how they can be cultivated without damaging ecosystems.

We will not deal with microalgae (such as *Chlorella*), single-celled organisms that function very differently,[2] and will also leave out the misnamed 'blue algae' such as spirulina, which were so-called until science recently discovered that they are actually not algae, but cyanobacteria.

The benefits of seaweed

Seaweed has learned to survive all over our planet. From opaline glaciers to sun-scorched lagoons, from salt-saturated seas to the fresh water of our rivers, it adapts to all environments, in all geographical locations.

The way it grows is exceptional. It needs only sunlight, salt water and nutrients from the ocean, does not take up space on land and grows more rapidly than tropical forests. It therefore absorbs much more carbon per acre than any terrestrial vegetation.

All of the world's sea forests cover an area equal to the entire landmass of the Amazon rainforest, but are at least twice as productive in terms of biomass.

Some types of seaweed have developed molecules with unique – and still largely unknown – properties that enable them to withstand extreme conditions. They live in symbiosis with an environment that is extraordinarily complex.

Others have learned to communicate with each other. Sometimes even outside their own species. If one of them is being eaten by a sea snail, it knows how to attract the sea snail's predators, such as crabs or fish, to defend itself.

Some, like plants, reproduce by cuttings. Others release female and male gametes, sometimes called spermatozoa, into the water.

Whether it's for our food, medicine, agriculture, livestock, cosmetics, textiles or substitutes for petrochemical industries, in all these areas, seaweed offers us sustainable and non-polluting solutions. Not to mention the species that could reverse the trend of global warming.

What can we do with seaweed?

Practically everything!

What do we currently know how to do with seaweed?

Practically nothing!

Is a seaweed revolution coming?

Phycology (or the science of algae) is practised by only a few scientists with limited budgets. These sea vegetables are teeming with qualities but remain relatively unknown.

How many do we know how to grow? A few handfuls at most. In most parts of the world, seaweed is gathered in the wild, opportunistically, and sometimes destructively.

12,000 years ago in the Middle East, what historians have called the first 'green revolution' took place. *Homo sapiens* ceased to be hunter-gatherers: we became farmers cultivating plants to feed our animals and our families. Since then, our civilizations have been built based on the domestication of the earth's resources. Theoretically, the cultivable ocean area is estimated to be 48 million square kilometres worldwide, taking into account solely nutrients and temperature.[3] We only cultivate 2,000. We are depleting our soil by cultivating 250,000 times more space on land than in our oceans, even though 70% of our planet is covered by the sea.

Meanwhile, at sea, we are still Stone Age hunter-gatherers.

But change is happening. In recent years, many local and international, private and public initiatives, NGOs and coalitions have become aware of the fragility and invaluable wealth of the oceans and seaweed. Furthermore, seaweed farming, which was non-existent at the start of the last century, spread to the Asian coasts to such an extent that, in 1960, half of the 2.2 million tonnes produced were cultivated. Today, 98% of the 35 million tonnes sold worldwide come from this cultivation,[4] which employs millions of people and helps feed almost two billion.

According to the FAO (the Food and Agriculture Organization of the United Nations), 'in 2019, the 34.7 million tonnes of world seaweed cultivation production for various food and non-food uses generated USD 14.7 billion first-sale value',[5] and the seaweed market continues to grow at a rapid rate. Currently, over 97% of global production comes from Asia. The rest of the world has not begun to cultivate these marine plants which, as mentioned, need only salt water and sunlight to grow.

Spreading the cultivation of seaweed throughout the world while taking care not to cause imbalance to ecosystems and recognizing the value of their compounds would offer us infinite potential for innovation and vastly increase the limits of our resources. A systemic approach that links the seas with our land-based societies could allow us to create truly regenerative agriculture.

Together, we could enter a new era, the result of a change as pivotal as the advent of agriculture in the Neolithic period.

An almost epoch-defining revolution...

Durvillaea antarctica ('New Zealand bull kelp') – a brown seaweed that measures between ten and fifteen metres and has the consistency of leather, it is found mainly in Chile where its stipes (stems) are used to make a traditional dish called *cochayuyo*. It is sold in tightly tied bundles at every market throughout the country.[6] This seaweed is also found in the rest of the South Pacific and was used by the Māori to make bags. It has a very strong holdfast. Given its environment, it must withstand forces equivalent to 1,100 kph winds on land. Like all brown algae, it reproduces sexually, and the algae produces spermatozoa and female gametes which use pheromones to attract them. It is named after the early nineteenth-century explorer Jules Dumont d'Urville, France's answer to England's Captain Cook.

I

FOOD: SEAWEED FOR FEEDING HUMANS

Monte Verde, Chile: America via the Seaweed Highway (14,000 Years Ago)

How did a small dam and a cave filled with seaweed at the foot of the Andes change our understanding of the settlement of the Americas?

When, in 1973 in Monte Verde, 500 kilometres south of Santiago and 30 kilometres inland, a farming family called the Barrías diverted the course of a small stream to help their oxen cross, they could not imagine the historical implications of this harmless act.

A year later, the erosion caused by this dam unearthed large bones belonging to mastodons.

In 1976, an archaeological expedition led by an American, Tom Dillehay, was sent there to carry out excavations.[7] This is when the scientific community established that *Homo sapiens* settled in America about 13,000 years ago via the 'Clovis Road', named after the city in New Mexico where remains of this 'first American civilization' were initially found.

So we knew for sure that the first Americans came from Asia around this period: they had crossed the lands that connected Siberia to Alaska in the area now called the Bering Strait. This region had just been freed from the ice and was not yet under water. The pioneers progressed to Central and South America over the following millennia. Numerous traces are still visible, bearing witness to their journey over hundreds of successive generations.[8] America was the final continent to be conquered by *Sapiens*, the first hominid to evolve there.

What Tom Dillehay found in Monte Verde was a water-logged bog that covered the floor of a perfectly preserved cave containing many remains. The high acidity of the soils and the peatbog had prevented bacterial degradation, and the organic residues were intact. About thirty people would have lived in this cave. A fairly imprecise initial dating put them at over 13,000 years old, which appeared to be a historical aberration. The team returned reports that were not published: no one agreed to validate them on the pretext that the models were erroneous, or the data was incorrect. The scientific community could not believe the theory of such an early settlement in America, at such a southern latitude. It simply could not be possible!

Years later, carbon-dating cleared up the mystery. The human remains were between 14,000 and 18,000 years old. This camp was therefore inhabited by human beings long before *Sapiens* arrived in America. This discovery was incomprehensible.

The old debate resurfaced between what the science says should be true and what is actually the case.

For scientists, there could be no other possible passage than the Bering Strait, but no previous human trace has ever been found on the tens of thousands of kilometres between there and Monte Verde. Plus, 14,000 years before our era, much of this land was frozen and unsuitable for the survival of a group of hunter-gatherers. The separation of the American continent from the rest of the world took place millions of years before the appearance of *Homo erectus*, and no human boat could have crossed the immense Pacific Ocean before our modern era. This whole hypothesis about such an early settlement was completely inconsistent. Unless aliens had come down from outer space and landed right there in Monte Verde...

But the expedition finally discovered conclusive evidence at the base of the cave: the dried remains of twenty-two different types of seaweed. Meticulously prepared, cut into small pieces, some of it chewed, it was obviously used for both food and medicine. Here, several hours on foot from the coast, the vegetation around the cave was lush and there was obviously no shortage of meat around. So why go to so much trouble to go and find seaweed?

This discovery gave rise to an entirely new theory about human settlement in America: the Kelp Highway.[9] Humans from Monte Verde would have simply followed the seaweed along the coast.

Gradually, the discovery of other equally ancient human traces along the coast, notably in present-day Oregon, confirmed this thesis. *Homo sapiens* settled in America well over 13,000 years ago. They travelled along the Pacific coast in small boats and over several generations. They followed this route because it was rich in resources that were necessary and familiar to them, notably the forests of giant kelp, known as *Macrocystis*.

Macrocystis can grow up to 60 metres in height and provide an ecosystem rich in shellfish, crustaceans, fish and other algae: all organisms that could easily have fed our distant ancestors.

Long before the gold rush, the American West had experienced a 'seaweed rush'!

This Kelp Highway danced along the Californian coast and then picked up again on the Peruvian and Chilean coasts. It allowed humans to progress south without going inland. Between the two continents, in the Central American region, they used the mangroves and the numerous red seaweeds found in these warm seas.

Why is there no trace of this descent along the coast? The reason is simple. The melting of the glaciers 12,000 years ago caused the water to rise by more than 100 metres, erasing almost all traces of these very first Americans who had no interest in venturing further into hostile lands. The remains of these seafaring pioneers have mostly been swallowed up by the ocean today.

But the presence of these twenty-two different types of seaweed in the Monte Verde cave, including *Porphyra* (*nori*), which is used to wrap our sushi, as well as *Durvillaea*, a traditional dish in Chile even today, demonstrates a great knowledge of marine plant resources. The efforts made to transport this food from the coast to the cave show its value at the time and suggest a complex trading network.

These facts also supported another – perhaps even more important – theory about who we are. It would be impossible for *Sapiens* to have developed such a disproportionately large brain compared to our body mass by evolving solely in the savannah. The genetic development of a neural network as sophisticated as ours requires specific micronutrients such as iodine and, above all, plenty of polyunsaturated omega-3 (EPA and DHA),[10] which is found in large quantities in seaweed and fatty fish but not on land.[11] It is therefore highly likely that when humans arrived in America they ate seaweed as well as the surrounding fish and crustaceans, as they had always done, or almost always...

It has now been established that, for several tens of thousands of years, the Heiltsuk, and other peoples of British Columbia in Canada, or Alaska, have celebrated the new year in the spring when the herring spawn on the seaweed. Thousands of eggs (up to 20,000 per female herring) are laid on the large kelp, which they cling to. These are collected and eaten, providing a considerable and vital quantity of proteins while respecting the balance of the ecosystems. The SOK (spawn on kelp) harvest is an institution, and in 1996 the Supreme Court of Canada confirmed the practice of this harvest as an essential and sustainable activity in these regions. The number of dishes is a clear testimony to these traditions originating from this first human migration.

Today, the consumption of seaweed as a food or medicine is still very common among the indigenous populations along this ancient 'Kelp Highway' from northern British Columbia in Canada to southern Punta Arenas in Chile.

So, whatever the pioneers of archaeology (mostly men...) may have thought, the resources necessary for the vertiginous development of our species did not come solely from men's mammoth-hunting prowess while women stayed in the caves with the kids! It is very likely that men, women and children collected these fragile little seaweeds and shells along the coast for food, and, in the process, they developed their brains in a spectacular way.

Monte Verde is currently the oldest site testifying to the intensive use of 'sea vegetables' by humans for food and medicine.

This cave tells us more about our own history than about the history of seaweed.

Our evolution on earth has been linked to seaweed for much longer than we thought possible.

A food challenge

The history of our food is undoubtedly the most incredibly rich cultural and social story of our species. Fire, flint, cooking, agriculture, breeding, bartering, storage, chilling, transport, markets, transformation, cultures, culinary traditions, industry, meals and social connections all represent significant discoveries and evolutions in our history. The order of priority for *Homo sapiens* has always been to feed our families, our clans, our tribes, and then our country. In today's globalized world, our challenge is to feed the planet. With healthy food.

This challenge is huge… Every day, there are nearly 250,000 more mouths to feed.[12] There will be almost 10 billion of us by 2050.

Where can we find these resources?

The equation is even more complex as the inhabitants of developing countries with rapid population growth aspire to consume more meat, the production of which uses more resources than for plant-based diets. Researchers have established that in the next fifty years, to feed the entire population, we will have to produce as much food on the planet as we have produced during the past 10,000 years.[13]

Today, more than 800 million people suffer from undernutrition and nearly half of the world's population is seriously nutrient-deficient. On the other hand, the abuse of processed, excessively fatty and sugary foods has become a major public health concern. In 2019, the EAT-Lancet Commission brought together thirty-seven of the world's top scientists and nutritionists from sixteen different countries to compile a global report on nutrition and its connection to our health and that of the planet. One of the conclusions of the report was that, to stay within the limits of possible global production, our diets needed to change radically. The report also stated that reducing our consumption of sugar and red meat by 50% could save around 10 million lives a year. Finally, it stated that an imbalanced diet (causing malnutrition, obesity, diabetes, etc.) currently poses 'a greater risk to morbidity and mortality than does unsafe sex, and alcohol, drug, and tobacco use combined'.[14]

At the same time, our planet no longer has much accessible and unused arable land. Urban sprawl continues to encroach on arable land. Cereal crop yields have not increased for decades and are even declining due to soil exhaustion and global warming. We are also the first generation to realize that our largely land-based food system is now one of the largest emitters of greenhouse gases. Our food system is one of the main contributors to droughts, soil depletion, loss of biodiversity, species extinction and pollution.

Even the most optimistic projections indicate that to feed the world in 2050 we will need 140% more water. We will also need to increase grain production by 50% and somehow add

14% more forest than we have today. In the process, we will emit around 80% more greenhouse gases, a far cry from our stated ambitions.[15] A very complex balance will therefore need to be found, and there will be no margin for error. To overcome these constraints, we have very few solutions on land.

But the oceans are a solution! They cover 70% of the planet and contribute only 2% of our calorie intake.[16] We only cultivate a third of our planet and dump our waste into the sea. As the primary link in the ocean food chain, seaweed could feed the planet if we cultivated it. At present, it is only North Asia that actually cultivates and consumes seaweed. Of the small amounts of seaweed produced outside Asia, 99% is still wild-harvested. Westerners have remained gatherers at sea.

In Asia, 99% of seaweed is cultivated. This cultivation began in the early twentieth century, but has only really developed over the past thirty years. 97.3% of the world's seaweed production comes from cultivation, and over 70% is for the food sector.[17] Much of it is exported to Europe, where it is incorporated, without us being aware of it, into many food products. We eat seaweed extracts every day and in large quantities.

Why is only Asia cultivating such a nutritious and abundant resource?

This question remains unanswered. Seaweed production on the Asian continent has increased from 4 million tonnes in 1990 to almost 35 million tonnes today. Over this period, these new sea vegetables therefore represent one of the highest growth rates in the world in food production.

Willem Brandenburg, an algae specialist from Wageningen University & Research in the Netherlands, calculated in 2015 that, by cultivating just 2% of the oceans, the protein requirements of the whole planet could be met without any additional intake of plant or animal protein.

Of course, we are not talking about eating seaweed and nothing else, but it's a matter of incorporating it more into our diets.

In Japan, seaweed accounts for 10% of daily nutritional intake, and this figure is the result of centuries of culinary traditions.[18] The current growth of the seaweed market would be sufficient to reach a production level that could meet this global demand by 2050. But this will require a change in eating habits and the cultivation of seaweed outside Asia.

If the food production potential of the oceans seems immense, it will be even greater when sea vegetables are integrated into a truly regenerative aquaculture system, including fish and shellfish farming. Considering that 70% of the world's population lives within 100 kilometres of the sea, this food then also becomes local.

The benefits of seaweed in our diet

All seaweeds are edible. Unlike terrestrial plants, there are no macroalgae that are toxic to humans, although not all of them have an appetizing flavour or texture.

The ones we see washed up on beaches are often unfit for human consumption because, detached from their substrate, they may already be in the decomposition stage.

Algae are nutritional bombs loaded with fibre and micronutrients, the contents of which vary from one species to another, and according to the season.

They are low in fat and contain a high amount of vitamins A, C and K, as well as iron, iodine, magnesium, phosphorus and zinc.

Some of them are the only plants to provide vitamin B12, which is necessary for the proper functioning of our brain. They also contain the valuable long-chain omega-3s (EPA and DHA) which are really important for the cardiovascular system and regulating cholesterol levels. These polyunsaturated fatty acids are rare; they are found only in fish oils and seaweed.

Seaweed also has incomparable benefits for functions related to digestion, blood circulation, the creation of muscle mass or the strengthening of bones. As far as protein is concerned, *Porphyra* (nori) and *Palmaria* (dulse) have a protein content of over 40% as dry weight, the same as soya, on which the protein intake of livestock farms is largely based, although it represents a huge environmental cost and contributes greatly to deforestation.[19]

To take another example, just 10 grams of the abundant sea lettuce (*Ulva*) is enough to meet our magnesium requirements. In this respect, it is much more natural and efficient than chocolate and contains no fat or sugar. It is also, and above all, rich in iron, calcium, potassium and vitamins A, C and B12.[20]

Several studies conducted in Japan by specialists in phycology have demonstrated the value of all these marine resources in the prevention of certain cancers of the breast, colon or prostate (we will come back to this in the chapter on medicine). It is important to point out that, unlike terrestrial plants, you don't need to eat a large volume of seaweed to reap its nutritional benefits. It is so nutritionally dense that adding a few grams a day to our plates in the form of dried flakes can ensure some of our daily requirements of essential nutrients.

The other nutritional value of seaweed lies in its structural strength, which preserves its contents even when it is dried out. Seaweed is capable of withstanding any aggression. Thanks to experience acquired over billions of years of adaptation, in the ocean, it can withstand extreme exposure to salt, cold, sun at low tide and persistent swells at high tide. Thus, unlike terrestrial plants, there are few nutritional differences between the fresh and dried versions of seaweed.

During the Covid-19 lockdown, Japan saw a massive upsurge in seaweed consumption. While the Europeans grabbed packets of pasta in the supermarkets, the Japanese rushed out to buy dried seaweed. It's not hard to imagine which part of the world has gained the most weight.

More generally speaking, this resource is nutritious and available all over the world. It retains its nutrients even when dehydrated, does not deteriorate over time, does not require a cold chain and is easy to transport without plastic packaging. All these characteristics are excellent news both for the populations of emerging countries and for the climate.

Drying these resources also has an advantage that has not yet been fully exploited. Indeed, by drying the seaweed, we recover water. Not surprising for an organism that spends its life in salt water! But the good news is that the water obtained from this process is fresh, demineralized water. Some seaweed-processing companies are considering the option of adding value to this by-product of seaweed, because demineralized water costs nothing to produce in this scenario, but can be sold wholesale for a few dozen cents. In Indonesia, a recent pilot project is exploring how to irrigate rice with water from seaweed, thus preventing massive investments into water-cleaning systems. Imagine this kind of co-production in arid regions, as well...

Drying seaweed is just one option for preserving it. The potential of fermentation is still largely unexplored, although Danish pioneers are actively working on this subject.

Seaweed fermentation is a technique that has been used in China for centuries, and more recently in Japan and Korea, by empirical means, with no reliable estimates of the benefits to date.

Current research seems to show that simple lactofermentation can make the worthwhile nutrients in seaweed, especially the proteins, much easier for our bodies to absorb. As with other foods, fermentation breaks down the cells in which the nutrients are locked. It 'predigests' them so as to make them available to our digestive system. This type of process also allows for a long shelf life and consumes less energy than drying. In addition, fermentation sometimes removes some of the strong flavours of seaweed and makes it more accessible to newcomers.

However, a still unknown number of parameters will influence the nutritional benefits of our sea vegetables. Indeed, the intake of a food is not equivalent to the total benefits of each of its compounds, but actually results from the combination of these compounds (the famous 'cocktail effect'). This is even more true for seaweed. Identifying the role of each of the relevant molecules and their interactions with other molecules in digestive environments is therefore very complex.

Other factors will be important for the quality of this nutritional intake: type of preparation, storage, processing, variety, geographical region, season and harvesting method. Genetic differences in our gut microbiota will also play a major role. Researchers still have work to do in terms of understanding and assessing how our bodies best assimilate the active compounds in seaweed.

Gastronomic innovation

Another challenge with seaweed is gastronomy, because on our side of the world we are as bad at growing it as we are at cooking it. But times are changing, and some top chefs are now becoming more interested in seaweed both for its texture and for its innovative flavours. Sea vegetables are uncharted territory for these culinary explorers.

Mauro Colagreco, an Italian-Argentine chef at the Mirazur restaurant in Menton (France), which has been awarded three Michelin stars and voted best restaurant in 2019,[21] even claims to use more than twenty different types of seaweed in his cooking. Trained in Japan, he plans to set up a seaweed farm near his restaurant so it can obtain the desired quality of ingredients every day. In Ireland and the US, many chefs are actively working with these new ingredients and incorporating them into their menus.

Unavailable commercially in Europe until a few years ago, local seaweed is now available in almost all organic food shops in fresh or dried form. Some large traditional supermarkets are also beginning to stock it on their shelves. Alongside this change, there are a growing number of websites[22] and specialized books[23] for seaweed-based recipes, from the simplest to the most sophisticated.

Seaweed contains multiple flavours and can even offer one that is lesser known to Western palates: umami, also known as the 'fifth flavour', in addition to salty, sweet, sour and bitter.

Umami means 'delicious taste' to the people of Japan, where it was only discovered through seaweed cooking in the twentieth century. An almost primal taste; breast milk is rich in umami. It is mainly due to a high presence of monosodium glutamate, an amino acid that is one of the most active neurotransmitters in our brain. Glutamate is also found in meat, fish and, in high doses, in Parmesan cheese, overcooked tomatoes and certain mushrooms.

The umami taste is enhanced by two other substances: inosinate and guanylate. The combination of these three compounds is only found naturally in seaweed, especially in kelp. Umami sends a signal to the brain to indicate the presence of proteins and triggers the secretion of saliva and digestive juices to facilitate their digestion.

In the early 2000s, scientists finally discovered the umami taste receptors in the taste buds. Incidentally, in 2010, the airline Lufthansa noted that the demand for tomato juice was surprisingly high on planes and even higher than the demand for beer. The German company sought to understand the reasons for this. Inflight, the pressure, air humidity and other factors modify our reception of smells and tastes. Perception of salty and sweet tastes is reduced by almost 30%; bitter, acidic and spicy tastes remain unchanged; while umami – which tomato juice is rich in – is increased tenfold, in particular by the noise of the plane which alters the functioning of the receptor on our tongue.

Seaweed, which is also very rich in umami, has been part of gastronomy in Japan, Korea and China for thousands of years. The finest varieties of seaweed are served to emperors, others are offered to deities in Shinto temples. The frequent presence of seaweed in Korean and Japanese cuisine is often cited as an explanation for the exceptional longevity in both countries. In Japan, this interest in seaweed is probably related to the historical need to feed a large number of people within a small land mass. In addition, Asia has been, and remains, influenced by traditional Chinese medicine, which attributes both preventive and therapeutic functions to food: an inclusive medicine that connects the individual to nature and the universe. It is impossible to understand Asian cuisine without mentioning the essential relationship it establishes between food, health and longevity.

The history and geography of seaweed in food

In Japan, 6 February is 'National Seaweed Day', a national holiday, and celebrating it has become a tradition. Up until the eighteenth century, taxes were paid with seaweed. There are Japanese poems dedicated to it. The five most popular types of seaweed in the Land of the Rising Sun are *nori, kombu, wakame, hijiki* and *mozuku.*

Nori, a red seaweed, is extremely rich in protein, vitamins and minerals. *Nori* means 'seaweed' in Japanese, but in common parlance it refers to a collection of red algae of the same species. *Porphyra,* to use its proper name, is a species of seaweed that has been consumed for centuries. There are historical traces of it as early as the eighth century, when it was eaten fresh. It was not until the Edo period (1603–1867) that the Japanese began to make it into dried sheets, to be consumed in a manner similar to how we do so today. It can be found as dried sheets wrapped around *maki* and other sushi, but also as granules, powder or flakes. It is to Japan what bread is to France.

Kombu is a brown seaweed whose name means 'happiness'. Naturally very high in umami, it is essential for the preparation of *dashi,* the basic broth of Japanese cuisine. It is found in all soups, but is also used as a condiment or in cooking dishes *en papillote.*

Wakame (*Undaria pinnatifida*), a brown seaweed, is also used in a great many dishes. In Japanese restaurants, it can be found floating in miso soup or served with sesame in the form of an almost fluorescent green salad (seaweed often changes colour when cooked).[24]

The last two seaweeds are less well known for various reasons.

Hijiki or *hiziki* (*Sargassum fusiforme*), a brown seaweed of the Sargassum family, is an ingredient in many traditional dishes in Japan. Low in vitamins, it is fourteen times richer in calcium than milk and contains more iron than most meats. Its omega-3 content is higher than that of any other seaweed. However, its high arsenic content, which is essentially not assimilable by the human body, makes its use rare in European countries, where legislation on the subject is unclear.

Mozuku (*Cladosiphon okamuranus*) is a brown seaweed formed of long filaments. It develops in symbiosis with other algae. Its name means 'seaweed stuck to it by growing on it'. It is only found in two areas of the world: Okinawa Island and the Ryukyu Islands in Japan. It can be eaten in soup or as an appetizer, or used in *tsukudani*, a traditional Japanese cooking method. It provides a unique combination of vitamins and other minerals. Its particular texture and taste make it a very popular food among connoisseurs. It is almost impossible to find outside of Japan.

South Korea shares much the same taste for seaweed as its Japanese neighbour. Its inhabitants consume less *mozuku*, but only because it does not grow there.

The country now produces four times more seaweed than Japan and appears to have made much more progress in modernizing the sector. Note in passing that the name

Kim, the most popular name in Korea, means 'seaweed'. *Kimbap* ('seaweed and rice') is the basis of the cuisine in the Land of the Morning Calm. The first written records of this dish date back to 57 BC. Koreans have been eating seaweed for centuries, yet it was only in the early 2000s that the government realized the incredible nutritional potential of seaweed and its health benefits. The 'Seaweed Day' programme, which was launched at that time, introduced sea vegetables in various forms into school canteens on a daily basis. South Korea has also specialized in the preparation of seaweed snacks, which it now exports worldwide. *Nori* is South Korea's second-largest aquaculture export after tuna,[25] and the industry provides tens of thousands of jobs. Koreans consume a lot of it but export even more, especially for the preparation of sushi.

While Japan remains both very traditional and measured in its production, essentially meeting its domestic demand, South Korea benefits from more favourable growing areas and has ambitious mechanization projects.

We also know, thanks to satellite images, that North Korea also cultivates and consumes large quantities of seaweed.

China remains the world's largest producer of seaweed, accounting for more than half of total global production.[26] However, its inhabitants consume a lot less of it than in Japan or Korea.

In the Pacific, from the islands of Polynesia to Hawaii, and among the native populations of the Chilean, Australian or New Zealand coasts, seaweed is also part of the culinary traditions.

Closer to home, the giant kelps present in northern Europe were essential for human settlement in Iceland and Greenland.

No land-based vegetables can grow there, while there is an abundance of high-quality seaweed.

A famous Icelandic saga from the thirteenth century tells the story of a father who decides to starve himself after his son dies. His daughter manages to trick him by making him eat a seaweed that she claims is deadly, but actually ends up giving her father back his taste for life.

Although seaweed isn't traditionally eaten in Scandinavia, it is said that the Vikings used to take a lot of dried seaweed on their ships to face their long journeys in the cold. This custom disappeared when the Vikings did, and without the link being established, long voyages to distant shores were almost impossible because of scurvy.

Most seafarers lost their teeth and suffered from haemorrhages.[27] As the days went on, they suffered from fever, lost the use of their limbs and experienced excruciating pain that couldn't be relieved by any medicine. According to Vasco da Gama, 120 out of his crew of 160 sailors died of scurvy. Magellan gives the figure of 247 deaths out of 265. In total, the 'plague of the sea' took the lives of several million men. The disease was so common that shipowners expected to lose half the crew on any voyage.

It was not until the discoveries of the famous Scottish physician James Lind that the link between scurvy and diet was established. While he did not discover that scurvy was the result of a vitamin C deficiency, he understood that the consumption of citrus fruits prevented the onset of the disease. So the ships were filled with oranges and lemons. And James Cook was able to land in Australia, New Zealand and then Antarctica.[28]

Long before Lind, the Vikings had intuitively understood this and carried seaweed on board. According to recent

studies,[29] 400 grams of seaweed would meet our daily require-
ment of vitamin C. For certain types of green seaweed, less
than 100 grams would suffice. It is a sad irony to think that
all these sailors died while floating over tonnes of the remedy
they needed.

No doubt there are still many regions of the world where
famines and other health crises could be avoided through a
better understanding of marine plant resources. The causes
are always multiple and complex, but the dramatic famine
which recently spread at high speed on Madagascar, an
island surrounded by an ocean rich in seaweed, unfortunately
seemed to mirror the fate of those sixteenth-century sailors.

In fact, on several occasions throughout history, seaweed
has solved food crises. In Ireland, during the Great Famine of
1848–50, people turned to seaweed, which has since retained
this image of being 'food of the poor'. In Russia, seaweed
appeared with the deportation of Korean populations from
Vladivostok to the centre of the country during the Soviet
period.[30] These kelp salads, which the Russians called *mor-
skaya kapusta*, were for years one of the cheapest and most
nutritious dishes in the country. Seaweed therefore contributed
to the food self-sufficiency of the Soviet Union during its
darkest hours, but was gradually forgotten after the fall of
the Berlin Wall and the opening up of borders with Western
countries.

In the north-west of Finistère, a poor region of France
with unproductive land, seaweed was cooked from the sixth
century onwards, when Irish monks brought over their traditional recipes for seaweed. These traditions appeared to have
been lost after the fifteenth century, although seaweed was
still used in the preparation of flans, as evidenced by many

nineteenth-century writings. Seaweed consumption picked up again in the late 1970s, first with the rise of macrobiotics and then by making its way onto restaurant menus in coastal areas.

Food risks or excessive regulations?

Despite seaweed's numerous benefits, as with all foods, some compounds can be harmful if consumed in excessive amounts.

The example of iodine, for which we have the most historical data, is a good illustration of the complexity, and the abuse of the precautionary principle which sometimes prevails. Remember: an excess of iodine is only dangerous for people with thyroid problems. Over the centuries, Asians have developed a specific microbiota capable of absorbing substances found in seaweed, including iodine. Studies have demonstrated the transfer of enzymes from marine bacteria into the gut of Japanese people, which are non-existent in the rest of the world.[31] These microbial enzymes enable the compounds in the seaweed to be assimilated much better and reduce the risk of overdose.

In addition, culinary traditions allow Asians to balance their daily intake. Seaweed is often eaten in the form of soup. Boiling kelp reduces its iodine content by 90%. They are also sometimes accompanied by vegetables, such as cabbage, which limits their absorption.

Unfortunately, in the West, the current regulations concerning seaweed products remain extremely disparate, poorly documented and very confusing.

For example, in Europe or the United States, the authorities have set very low iodine tolerance levels,[32] which makes it impossible to market many species of seaweed produced in these countries and therefore limits the development of the sector.

This tolerance threshold is difficult to understand if we consider the fact that a significant part of our population is deficient in iodine.

Symptoms of this deficiency usually result in a feeling of fatigue, depression, weight gain and reduced immune capacity. So it seems a shame to prevent producers from marketing products that have a very beneficial nutritional content.

The stance taken by our health authorities seems even more astonishing when we realize that the permitted tolerance levels for iodine content in seaweed in countries where it is eaten every day, such as Japan and Korea, are sometimes a hundred times higher than in the West.

Culinary traditions and microbiotic differences cannot explain such a discrepancy, which is probably due rather to a lack of knowledge or to excessive precaution.

The illogicality does not end there. Under current international trade agreements, seaweed imported from Asia that complies with local legislation can de facto be imported into Europe and elsewhere. So, in theory, it would be possible for a European producer to put his non-compliant seaweed production on sale on the European market by exporting it to Japan and then reimporting it…

The other health risk to consider concerns the content of heavy metals such as cadmium, lead or aluminium. These marine plants absorb and store minerals floating in the ocean. If they are harvested in highly polluted areas, eating them can be dangerous for our health. Strict regulations exist everywhere to prevent the sale of sea vegetables with excessive heavy metal content. But here again, there is no consistency when it comes to the regulations. It seems incomprehensible and scientifically illogical that the tolerance levels for cadmium in seaweed in Europe are often much lower than the permitted levels for cadmium in seafood, fish or even potatoes.

Another prime example concerns arsenic. Although the word probably doesn't make your tummy rumble, it should be noted that this element can come in organic or inorganic form. Only inorganic arsenic is assimilable and, in certain doses, will be toxic to our body.

Organic arsenic, which is also present in many other products such as fish or mushrooms, cannot be assimilated and will be completely eliminated through natural channels. It therefore presents no danger. While certain seaweeds do contain arsenic, it is only in organic form and therefore harmless. The small inorganic part found in some seaweeds such as *hijiki* in Japan can be removed by a long boiling process. Alas, there is often major confusion over this, and many products are banned from the market simply for having the word 'arsenic' in the list of their compounds.

The disparity of seaweed regulations around the world tends to slow down their use by major international food brands, despite the nutritional benefits of these plants and the growing demand for plant-based diets.

We can only hope that new scientific studies carried out in conjunction with standardization and education of the relevant authorities will soon establish the benefits of seaweed for humans. Furthermore, seaweed producers, like other corporations, need to come together to set consistent standards around the world, to protect consumers while avoiding abuse of the precautionary principle.

It is precisely with this in mind, but also in response to the massive fragmentation of information and the total lack of collaboration between market participants, that the first global coalition of seaweed stakeholders was launched in early 2021. The aim of this coalition, created in collaboration with the CNRS (French National Center for Scientific Research), with the support of the United Nations Global Compact and the Lloyd's Register Foundation,[33] is to promote knowledge sharing and to set up working groups capable of providing the expertise needed to develop fair and equitable international safety rules for seaweed products (food or otherwise).

The coalition is also working on environmental and social standards related to the cultivation of these marine plants, in consultation with the major international institutions, with the goal of making it a modern, ethical industry. It therefore aims to promote the development of comprehensive regulations, cooperation between stakeholders, as well as investment in the seaweed sector.

To be clear, it is neither possible nor desirable to base our diet entirely on seaweed. We need diversity, and sea vegetables should be on our plates in addition to other products.

However, increasing the demand for seaweed for food purposes remains the cornerstone of developing the seaweed sector, above any other use or processing. Experience in the marine plant sector shows us that a mass market can only work if human food is at the base.

The major investments required for fundamental research, as well as for the implementation of new technologies or international regulations, can only be granted to meet a significant development in food demand.

As we saw earlier, the only truly successful market for sea-weed is in north-east Asia, where it is widely cultivated and even more widely cooked.

We already know how decisive and political our consumer choices are, especially regarding the food we consume.

We are all agents of this change. Thanks to seaweed and marine resources, we have the opportunity to become the first generation on this planet to feed the entire world population in a healthy and sustainable way.

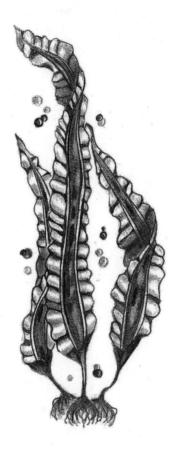

Saccharina latissima ('sugar kelp', 'sea belt' or 'Devil's apron') – a species of *kombu*, it is a sucrose-rich brown seaweed measuring one to three metres in length when fully grown. It is characterized by its golden colour and waffle-like fronds which, in water, sometimes gives it the appearance of a crocodile. Found throughout the North Atlantic, it was one of the first seaweeds to be cultivated in Europe and America. Often used to season dishes or for *en papillote* recipes, it has also been widely used for centuries for animal feed and is also consumed by marine fauna. It is rich in alginate and contains acrylic acid, which shows a broad spectrum of antibacterial activity.

2

FARMING AND LIVESTOCK: SEAWEED FOR FEEDING ANIMALS AND IMPROVING AGRICULTURAL PRODUCTION

North Ronaldsay Dyke in the Orkney Islands, Scotland, 1832

The Orkneys, a tiny archipelago lost in the ocean on the northern borders of Europe, in Scotland.

A little further north is North Ronaldsay, the last port of call before the (for now) endless ice caps of the Arctic. An islet of less than three square kilometres, floating in the middle of immense waves that lap its shores and deposit large swathes of brown.

Around the perimeter of the island, there is a very old dry-stone wall, known as a dyke. Twenty kilometres long, it encircles the whole island without the slightest chink in its armour. Beyond the dyke there is nothing but a vast expanse of seawater and yet, to this day, the wall is guarded day and night by a warden who diligently keeps watch over the other side of the wall, and protects the dyke from erosion over time.

The structure was built by the islanders in 1832. At that time there were almost 500 of them, ten times as many as there are now. North Ronaldsay is a remote, barely accessible place, far from the lights of the big cities. The soil is barren, the climate harsh, the island battered by driving, icy rain during the endless dark winter nights. This tiny rock now has only around fifty inhabitants… what terrible danger could they need protecting from? The dyke is the longest of its kind in the world but is less than two metres high. It does not exactly seem like it could repel an invader.

In fact, the dyke wasn't built to protect the inhabitants of the island from external dangers, but rather to keep the creatures who live on the other side of the wall safe.

Who are the creatures living on this narrow strip of land between the dyke and the ocean?

They arrived in Orkney at least 10,000 years ago and have survived every hardship that nature and history have inflicted on the inhabitants of that hostile place. This population is now extremely coveted and can be found only in this part of the world. They feed almost exclusively on the marine plants in abundance there at high tide…

They are the legendary, peaceful sheep of North Ronaldsay, a rare breed, highly prized for their wool and exceptional meat, and enjoyed by the British Royal Family on special occasions. Research carried out on dental remains of distant ancestors of these animals shows that they have been feeding (at least partially) on seaweed for over 6,000 years. Over time, their gut microbiota learned to digest it, to tolerate its high salinity levels and to extract its levels of copper to meet the sheep's physiological needs. They are not the only mammals to do this: Shetland ponies and red deer are also known to graze on seaweed regularly.

North Ronaldsay sheep have become unique in that they can now feed solely on seaweed.

What's more, unlike other sheep which eat during the day and ruminate at night, North Ronaldsay sheep have adapted their circadian rhythm to that of the waves. They therefore graze twice a day at low tide and then ruminate twice a day at high tide. This rhythm prevents them from finding themselves without food or being trapped by the waters at high tide.

So why the dyke?

In the early nineteenth century, the human population of the island grew rapidly. At the time, the Kingdom of Spain, which supplied Europe with potash – an essential element for the production of soap and fertilizers – was too busy dealing with the attacks of Napoleon I to meet the growing needs of the surrounding countries. To overcome this critical shortage of potash, the British Empire decided to use sodium carbonate, which could be produced from the abundant supply of seaweed, in particular on the island of North Ronaldsay.

Once the war was over, the trade in cheaper and more efficient Spanish potash was re-established. This left the Scottish islanders without resources.

Starving, they decided to develop agriculture. Alas, the island's endemic sheep, which grazed away like mad on their meagre crops, scuppered their efforts. The land, with its low yields, was too precious to be shared with these livestock.

On the other hand, the inhabitants could not bring themselves to be parted from their faithful companions, swathed in such beautiful wool.

So the people of North Ronaldsay took a wild gamble. In order to keep their sheep and preserve their land, they decided to enclose the sheep behind a dyke, between the land and water. They knew these sheep loved seaweed and bet that their adaptable nature would lead them to feed exclusively on it.

The experiment was daring because no other land animal, with the exception of a Galapagos iguana, can survive solely on eating marine plants.

But the gamble paid off: the sheep survived, and the excellence of their meat and their wool made this small island famous.

As the seaweed thrown up by the storms were even more abundant in winter than in summer, the animals gained weight mainly in winter.

It has also been shown that their digestion was improved by this diet and that their methane emissions were consequently much lower than those of other sheep.[34] This point is far from trivial, knowing the dramatic contribution of methane emissions from ruminants to climate change.

Gradually, it became impossible to let the sheep return to the other side of the wall because, in addition to the risk of interbreeding with other breeds, letting them graze too often on terrestrial plants would have exposed them to a copper content that their altered metabolism could no longer tolerate.

And so, the 2,500 North Ronaldsay sheep are still regularly moved to land-based pens at certain breeding times, but under strict supervision.

And even in this small enclosure on land, they prefer to eat seaweed rather than grass!

A few years ago, the Orkney Sheep Foundation decided to continue to protect these sheep while rebuilding and repairing the dyke, since maintaining it was becoming very difficult for the few families still living on the island.

The Foundation has therefore created a post of shepherd,

who more often than not is a shepherdess, tasked with watching over the flock and reinforcing the dyke.

The efforts of the shepherdess are complemented by those of the visitors who come to the North Ronaldsay Sheep Festival every year. The sheep are a source of immense pride and a major tourist attraction for the Orkneys. The quality of the resulting products is also a great economic asset for the region.

More than a story about sheep, this is a story about seaweed.

Animals trapped by the tides have adapted, regaining their natural connection with the marine world. These sheep have gone back in time to resurrect the past locked in their mammalian cells.

For 6,000 years, they have been proving the nutritional benefits of seaweed and demonstrating every day that a more sustainable alternative to Brazilian soya or meat and bone-meal is still possible for livestock.

Feeding livestock and improving agriculture

Enthusiastic advocates of the merits of sea vegetables are very often met with deep scepticism. The unbelievers refuse to move away even momentarily from the taste of a grilled prime rib or the nostalgic pleasure of their childhood shepherd's pie, for what they see as a slimy, ugly, smelly organism.

If we overlook the reductive aspect of this conception of seaweed cuisine and the environmental cost of meat production to our planet, this point of view is completely acceptable, and is likely to remain a reality for some time yet, for a significant part of the population in the West.

We could nevertheless make this argument more nuanced by recalling that eating habits change more quickly than we think and are often linked to the latest fads and social norms. In the Middle Ages, 'productive' animals were rarely on the menu, as people preferred to keep chickens for eggs, cows for milk, sheep for wool and oxen for agricultural work.

The meat consumption of the nobility was limited to game, the only meat really worthy of being eaten, while the peasants were content with wild rabbits, small birds and, for the less poor, pigs for celebratory feasts. Even in the early twentieth century, lobster was only eaten by prisoners, orphans and cats!

Not to mention raw fish, a dish that was inconceivable in our grandparents' day.

We could also say to those who are resistant to eating seaweed that they already eat it every day, perhaps several times, without even knowing. Seaweed is widely used in the food industry as a texturizing, gelling or thickening agent.

Despite all these arguments, seaweed may take a few years to become part of Western countries' gastronomy...

The best answer would be to keep in mind that sea vegetables have the potential to make the production of land plants, fish and meat more sustainable. Indeed, these 'superfoods' are as beneficial to animals as they are for humans.

It would be possible to build sustainable aquaculture, to better feed our land-based farms and to naturally stimulate plants using spreading, which has been practiced for centuries in coastal regions.

Seaweed, especially kelp, has always been present in animal feed. Throughout Northern Europe, it has been an integral part of the fodder given to livestock for many centuries, and the vernacular names of certain seaweeds reflect this: 'cow seaweed' in Brittany, 'horse seaweed' or 'pig seaweed' in Scandinavia.

Permaculture at sea: towards regenerative aquaculture

The first case to consider, logically, is seafood.

Seaweed, in its natural state, is the first level of the aquatic food chain. The exudates (pieces) that break off and float in the ocean represent up to 50% of the total algal biomass. This feeds plankton and 'filter feeders' (clams, krill, sponges...), which then serve as food for the rest of the aquatic food chain.

Aquaculture is now the fastest-growing sector of the food industry. Its development has been exponential in recent years, and this trend is expected to increase in the coming years.

This type of cultivation has many advantages for providing nutrient-rich, sustainable food for the inhabitants of our planet. Here again – and even though the imbalance is less than for seaweed – it is only Asian countries that have really developed this system in an ambitious way. Aquaculture production in China, India, Indonesia, Vietnam and Bangladesh accounts for 80% of the world's seafood culture.

On this subject, it should be noted that, although the word 'aquaculture' refers in our minds to the farming of fish in the ocean, sea aquaculture (or 'mariculture') represents only 55% of total aquaculture production in terms of volume.[35]

Within this mariculture, seaweed dominates. It accounts for more than half of the volume cultivated (51%), while molluscs account for less than a third (30%), and fish and shellfish share the remainder.

The distribution is quite different if we consider these resources in terms of their value, with carnivorous fish at the top of the food chain (salmon, trout, cod, etc.) obviously being much more expensive than seaweed.

Fish remains the most popular sea product in our food traditions.

With a fivefold increase since 1990, farmed fish has only just overtaken fish from fisheries in terms of volume. In addition to the fact that it does not require cultivable land, fish is a product that is rich in the nutrients our bodies need, and its conversion rate (ratio of dry weight of feed distributed to production gain obtained) is much better than that of any land animal.

Indeed, by not consuming energy either to heat up its body (fish are heterothermic), or to combat gravity, fish will transform one kilo of food into 900 grams of body mass, whereas oxen will produce less than 150 grams.[36]

This productivity saves a lot of resources and limits carbon emissions.

Aquaculture also avoids the pitfalls of fishing in terms of the risk of depleting resources.

Despite this, in the West, it is perceived – very often with good reason – to be extremely harmful to the environment.

First of all, fish feed, which is purely extractive, comes either from land-based resources (wheat, soya, animal meat, oil, etc.), which contributes to deforestation and soil depletion, or from fishing products (anchovies or krill), whose over-exploitation leads to the disappearance of species and very serious ecological imbalances in the ocean.

This situation is the result of the sector developing in the opposite direction to the food chain.

12,000 years ago, when humans emerged from the Stone Age

to progress from hunter-gatherers to farmers and breeders, it is likely that we began by growing cereals for food.

In a secondary phase, we probably used these cereals to feed animals and to develop breeding activities in order to obtain milk, wool or eggs. In addition, much more recently, as our yield in cereal production has become very high and our mastery of livestock production significant, we have increased the number of animals bred mostly for eating.

As for the ocean, modern humans have not had the patience to learn how to domesticate plants to rebuild a food chain capable of feeding fish. Aware that we are reaching our limits for land-based food resources as well as fisheries, we have urgently decided to relieve these sectors by turning to breeding fish fed by land-based food resources or fisheries!

These intensive fish farms, grouped together in small areas, often close to the coast, represent major sources of pollution for the ocean. Fish faeces and unabsorbed feed residues from marine farms concentrate in these waters and disrupt the aquatic environment, contributing to the creation of areas with no oxygen (the degradation of this waste by bacteria consumes all the oxygen). Their waste deteriorates the quality of the water and destroys the habitats of marine flora and fauna.

This mode of production makes the same mistake of land-based agriculture by relying almost exclusively on monoculture models that deplete ecosystems. Aquaculture is therefore becoming a serious environmental nuisance, despite its potential to better feed the planet.

However, another regenerative and less environmentally damaging model is possible. It is called 'IMTA' (Integrated Multi-Trophic Aquaculture), or permaculture at sea, or '3D Farming' in the US.

The principle is to breed fish alongside seaweed and invertebrate cultures so as to recreate, at sea, similar regenerative ecosystems to those that exist naturally. The polluting organic waste from fish or their food is therefore recycled by molluscs (mussels, scallops, etc.) while the inorganic waste from fish and molluscs is absorbed by the seaweed. To complete the circular nature of this integrated ecosystem, it is possible to line the seabed with sea cucumbers, sea urchins or starfish to absorb the residual waste from other productions and prevent it from becoming concentrated in ocean sediments.

This fertile combination of ecosystem benefits provides a new reservoir for marine biodiversity while optimizing space and preserving the balance of the seabed. After harvesting, each species can be used individually and sold in its own sector to generate income.

This system essentially sets up a huge natural waste recycling plant in the ocean, which repairs the environment instead of degrading it. This revolutionary concept is actually nothing new, as the earliest treatises mentioning this type of practice in China date back 4,000 years. At that time, freshwater fish farms were combined with seagrass and shellfish farms. However, this system has seen a resurgence of interest in recent years due to the combination of an awareness of the damage caused by fish farming and a growing interest in seaweed farming (also known as algaculture).

However, to date, no large-scale commercial system has been implemented.

It must also be recognized that, while the model is inspiring, the technical and regulatory constraints are onerous. The different species must be well selected and their cycles carefully studied. This is because seaweed generally takes up nutrients seasonally, whereas fish produce the same amount of waste all year round. This model is also complex for producers who struggle to sell their various products on independent markets that function in different ways. However, there are more and more initiatives out there.

In France and Norway, the concept is being trialled. Eastern Canada has also developed interesting skills and pilot sites. The major international salmon-farming brands are actively working on the subject in order to make their production sustainable. Other projects aim to use certain kinds of containers filled with seaweed that would be connected to fish farms via large pipes, to recycle aquaculture waste while producing valuable resources.

In Portugal, Algaplus, a pioneering company specializing in seaweed and the only one currently capable of producing Atlantic nori, has managed to set up this system for growing seaweed, in sea bass farms.

In Asia, things are developing in reverse. It starts with seaweed, which is already intensively cultivated, and uses its waste to feed fish and shellfish.

Simpler integrated systems that use seaweed to feed molluscs or crustaceans are already quite widely developed around the world, but fish are still too rarely integrated into them.

In India, a major shrimp producer, seaweed has recently been introduced into shellfish farming in order to limit the serious damage they cause to the environment. Shrimp farms are responsible for about 30% of the destruction of mangroves.[37] Mangroves are, however, very rich ecological niches: they allow marine fauna to reproduce and absorb a very large amount of carbon.

The destruction caused by intensive shrimp farming is largely due to significant releases of nutrients (food, manure), which promote the development of organisms that are harmful to mangroves.

Growing seaweed would absorb these nutrients and therefore protect the mangroves, while creating a valuable additional resource for fish farmers.

A regenerative system of aquatic permaculture that reproduces natural life cycles is the future of a sustainable aquaculture, one which would make it possible to feed billions of people on nutrient-rich food without degrading the environment.

Integrating the cultivation of these species could then be taken a step further by developing the rest of the food chain, so that fish can be fed without the need to seek land-based resources or to resort to fisheries. It is already possible to use seaweed as a food supplement for fish feed. Seaweed is certainly not sufficient to feed carnivorous species, but using it as an additive makes it possible to promote the growth of fish while significantly strengthening their immune systems in order to avoid – or at least reduce – the use of antibiotics.

In the Nordic countries, highly specific seaweed extracts are being studied with a view to developing an immunostimulant to prevent fish from being parasitized by 'sea lice'. This is an extremely interesting development, bearing in mind the economic disasters caused by this scourge, which threatens a good part of the production of salmon throughout the world.

In the 2000s, this crisis forced producers to dump hundreds of litres of pesticide into the fjords, which in the end did much more harm to the surrounding ecosystems.

Seaweed for more sustainable land-based farming

Seaweed can also be beneficial for terrestrial animals. In his text written in 50 BC, *Commentaries*, Julius Caesar recounts how he won a decisive battle in northern Africa by feeding his starving horses on dried seaweed, for lack of grain.

For centuries, seaweed has been used to feed animals, and farmers have regularly testified to its positive effects.

Nowadays, thanks to scientific advances, we can better characterize the benefits of seaweed for animals, and extract specific compounds from them to enhance the expected effects.

The largest seaweed farm in the Faroe Islands, which is ahead of the game, feeds its entire yield to pigs in northern Europe. The seaweed is fermented with milk yeast so it is better assimilated by the animals. Studies seem to show that this consumption increases the growth rate of pigs by more than 10%, while reducing their mortality rate and improving the lactation of sows to feed piglets.[38]

Both in fish and in certain terrestrial animals, seaweed extracts are capable of acting on the intestinal microbiota to support natural defences or to reinforce the effectiveness of a vaccine. In France, chicken farms, fish farms and egg production facilities are now able to do without antibiotics, by using seaweed as a feed supplement. A professional association called 'Merci les algues' brings together scientists, producers, processors and distributors to test and promote the use of seaweed in animal feed and agriculture. For example, since 1 March 2020, one of the French market leaders has been marketing free-range eggs without antibiotic treatment, thanks to the incorporation of solutions based on local green or red seaweed into their feed via drinking water.

In all intensive farms, the risk of contagion is high and animals are often treated with antibiotics to prevent disease. These antibiotics end up on our plates and account for 80% of the antibiotics we ingest. This massive and indirect consumption promotes what is known as antibiotic resistance. When exposed to an antibiotic, bacteria adapt and become resistant to it.

In the last century, the discovery of anti-infectives, and antibiotics in particular, was an extraordinary advance which considerably lengthened life expectancy. Unfortunately, today, the molecules used are less and less effective, finding new ones is a complex process and the phenomenon of antibiotic resistance is already responsible for around 700,000 deaths per year worldwide.

According to a UK study in 2016, infections due to resistant infectious agents could once again become one of the leading causes of death in the world by 2050, causing up to 10 million deaths per year.[39]

Using the properties of seaweed to improve immune defences and reduce the use of antibiotics in animal husbandry is no mean feat, but it still holds immense promise for the health of future generations.

The benefits of seaweed don't end there. Other seaweed-based products limit dysfunctions of the digestive system or kidney problems and improve, among other things, the quality of milk and the longevity of dairy cows.

In general, the consumption of seaweed by animals improves the quality of their meat, as in the case of our Ronaldsay sheep, and increases the production of their resources (milk, eggs, wool, etc.) This improved productivity is also often linked to an improvement in the well-being of the animal, which is sick less often, and has better digestion. There is therefore also an ethical aspect to the proposal for food supplements derived from seaweed.

As already mentioned, seaweed is also very rich in protein, although more research is needed into improving its bioavailability to animal organisms. The fact remains that currently the protein intake for our farms comes mainly from soya. More than 85% of the soya produced in the world is used to feed animals. Soya has a high vegetable protein content and is a major part of our animals' feed, as it encourages the development of muscle and therefore meat.

Most of this soya comes from Brazil, where it is the main cause of the dramatic acceleration in deforestation in the Amazon.

From there to the animal troughs, it has to cross the entire planet: a catastrophic carbon footprint.

Finally, although two thirds of the soya produced is already genetically modified, it requires large-scale use of pesticides and fertilizers, which contribute to the pollution of soils and rivers, in particular in the Amazon in South America.

Looking for – even partial – alternatives to soya in order to meet the protein needs of our farms seems to be an excellent starting point to help our planet. The protein content of soybeans in dry matter can be up to 35%. For some red seaweeds, this figure can be as high as 45%.

The other advantage of seaweed over soya is its content of long-chain omega-3 fatty acids (EPA and DHA). Omega-3 and omega-6 fatty acids are vital compounds for our bodies, but our bodies are unable to synthesize them. Soya is naturally very low in omega-3, which generates a serious imbalance between omega-3 (anti-inflammatory) and omega-6 (inflammatory) in animals.[40] This imbalance is logically then found in consumers of these animals, which explains why 90% of the French population suffers from an omega-3 deficiency.

From the time when *Homo sapiens* appeared on earth up until the 1950s, the ratio of omega-3 to omega-6 was on an equal footing. An imbalance has arisen with the development of modern food systems that are based on high – and often indirect – consumption of soybeans and vegetable oils from rapeseed or maize, in the form of processed products. Studies clearly show that a level of omega-6 five times higher than omega-3 creates predispositions for cardiovascular diseases, cancer and inflammatory and autoimmune diseases.[41] The imbalances observed in today's Western population indicate levels of omega-6 more than sixteen times higher than levels of omega-3!

Omega-3-based treatments are shown to be highly effective in reducing the impact of the above-mentioned diseases. They are also effective against mental disorders such as depression, Alzheimer's disease and bipolar disorder.[42] In order to rebalance this ratio between our fatty acids, it is necessary to favour products rich in omega-3s. While certain plants such as flax or chia seeds are rich in omega-3s, these types of short-chain fatty acids (ALA) are not easily assimilated by our organisms. The only long-chain omega-3s (EPA or DHA) that really benefit us are found in seaweed[43] or fish oils.

As we saw earlier, these fatty acids are precisely the ones that allowed us to increase our brain mass in order to become the humans we are today. More research is needed to understand how to better benefit from them. If we don't consume them directly by eating seaweed, it might be a good idea to feed it to our animals so that we can indirectly reap the benefits.

Seaweed and methane emissions

Certain types of seaweed used to feed animals can help reduce our emissions of very harmful greenhouse gases, such as methane. The IPCC (Intergovernmental Panel on Climate Change) reminds us that agriculture currently accounts for 24% of greenhouse gas emissions.[44] The sector, which has seen a sharp rise in emissions, is in second place behind energy production (which has seen a 25% decrease due to renewable energies). Methane is the second-largest contributor to global warming, accounting for about 20% of global emissions. This gas is thirty times more harmful to the atmosphere than carbon dioxide and contributes eighty times more to global warming over a period of twenty years.

A large proportion of methane emissions come from the gaseous emissions of ruminants, which, according to the FAO,[45] emit 3.5 gigatonnes of methane every year, which is as much as cars and other means of individual transport.[46]

However, very recent in vitro studies have shown that incorporating 0.2% of a species of red seaweed named *Asparagopsis* (red sea plume or *limu kohu*) into the diets of ruminants would make it possible to improve the animals' digestion and well-being, but above all would cause a reduction of more than 80% of their methane emissions.[47]

The reducing effect comes from the presence of elements in this red seaweed that can modify the composition of the intestinal flora, to suppress the bacteria responsible for the production of methane. If it were fed to all our ruminants in such a small proportion, it would be equivalent, in terms of greenhouse gas emissions, to stopping all the cars on the planet immediately.

Meat accounts for a quarter of global methane emissions, and the United Nations Environment Programme (UNEP) stated in May 2021 that 'cutting methane emissions is the best way to slow climate change over the next twenty-five years'. This statement is based on the fact that carbon dioxide remains in the atmosphere for hundreds or even thousands of years. These timelines imply that even if emissions were immediately and drastically reduced, the impact on the climate would not be effective until the end of the century. However, it only takes about a decade for methane to decompose. So reducing methane emissions today would have a huge short-term impact on climate change.

Methane is also a major contributor to air pollution and the UNEP report states that a 45% reduction in methane emissions worldwide would prevent 260,000 premature deaths, 775,000 asthma-related hospitalizations and 25 million tonnes of crop losses per year. The subject is therefore far from trivial.[48]

It is advisable to remain very cautious about these assertions concerning *Asparagopsis* because we lack hindsight on these data and the long-term benefit for animals is not yet confirmed. The bromoform contained in this red seaweed is still a carcinogenic substance, and the long-term health of the animals must be ensured. The efficacy of the process in vivo needs to be replicated on a large scale, and to ensure that, over time, neither the productivity nor the appearance, taste or smell of milk and meat are altered in these animals. Finally, we need ensure the sustainability of this phenomenon over the years by checking that the methane-producing bacteria do not regenerate.

The other limiting factor has to do with supply. This variety of seaweed presents a challenge because as a wild resource it only grows in a few places and its large-scale cultivation is still in an embryonic state. An Australian company based in Tasmania, where this little red seaweed is very common, launched the first large-scale experiments in 2020. It has secured 5,000 hectares (!) in a concession agreement with the government and has managed to domesticate *Asparagopsis* and seven other endemic seaweed species. It is pursuing its tests, in particular with New Zealand's cooperative dairy company Fonterra, the undisputed world leader in dairy production.

The initial results are very positive, and the digestive energy not utilized to produce methane seems to be used to produce more meat.

As a result, ruminants have lower methane emissions and a 20% increase in the growth rate of their meat. This point is all the more important, as there is currently no programme to encourage farmers to reduce methane emissions. Being able to rely on positive collateral effects in productivity is therefore essential.

There, for example, the extra cost to producers of using this natural, almost methane-free and growth-enhancing seaweed is around a dollar per day per cow. This expense is significant but becomes worthwhile when you consider that the cost of traditional growth hormones is about double that amount.

The Tasmanian company's production could eventually cover the needs of half of Australia's cows.

Millions of dollars are being invested in other companies with the same objective, and the world's largest seaweed-growing factory is currently being built in Sweden to produce the same seaweed in an indoor environment.

But this little red seaweed is not the only option, and there are high hopes for types of brown seaweed already widely used for animal feed for centuries, such as *Saccharina latissima* (*kombu*).

These would make it possible to generate, after lactic fermentation, a substantial reduction in methane emissions by ruminants. Studies are currently underway. Even if the result turns out to be lower than those obtained with *Asparagopsis*, the opportunity is still remarkable, because these types of large kelp are easy to cultivate everywhere around the world: they grow quickly and animals have eaten them for millennia without any side effects...

Especially since the use of fermented *kombu* also seems to improve the growth of animals.

As always, a great deal of research and testing will be needed to properly master these new techniques and bring together a set of practices that are more respectful of natural balances for animals and the environment. But demonstrating their long-term effects on methane production would be excellent news, probably decisive in enabling us not to exceed 1.5°C of global warming according to the objectives set by the Paris Agreement at COP 21 in 2015.

Seaweed as a natural biostimulant for plants

As well as providing interesting solutions for our farm animals, we must also remember that seaweed offers solutions for the terrestrial plants that feed us. As they grow, marine plants, bathed in nutrient-rich seawater, are loaded with mineral elements (nitrogen, potassium, calcium, magnesium, copper) to produce amino acids, hormones and complex sugars such as alginate.

Once spread on the fields, they release these substances which stimulate natural processes, both root and aerial, to improve nutrient uptake and tolerance to abiotic stresses (cold, salinity or drought).

In this case, seaweed becomes a 'physioactivator': it improves plant performance and replaces chemical fertilizers or fertilizers with side effects, or even genetic modifications. Studies have shown that seaweed allows plants to absorb up to 20% more nutrients.[49] Vines, cereals, soybeans, cotton, potatoes and even fruits can be treated with seaweed, in order to replace fungicides, pesticides, herbicides and repellents at the same time.

Trials are underway into combating *Septoria tritici* in wheat, vine downy mildew and cucumber powdery mildew.

Nearly fifty companies around the world manufacture biostimulants for plants out of seaweed. A company in France has even developed a vaccine for cereals, apples and strawberries from *Laminaria digitata*, a brown seaweed that is very abundant in Brittany. These vaccines are approved in a dozen countries and recognized as active against many pathogens.

These products stimulate the immune defences of plants so naturally that they encounter unexpected difficulties. Thus, the lack of a quantified estimate of the residues of these algal molecules added to terrestrial plants poses enormous problems when it comes to introducing them to the market in some countries. These molecules either occur naturally in the treated plants or involve variable and uncalibrated molecules, which do not comply with the usual artificial and synthetic standards. Administrations are therefore struggling to authorize these products, which are too far removed from highly stable and easily detectable synthetic chemicals. So a producer of these seaweed-based biostimulants faced Kafkaesque complexities – due to an inability to quantify the residues – when trying to validate the safety of a product that was nevertheless organic and natural.

In soils, seaweed extract also accelerates the decomposition of organic matter, increases the bacterial population or the activity of earthworms and improves water-retention capacity through the alginic acids or other polysaccharides in red and green seaweeds.

The use of seaweed in agriculture dates back hundreds of years. On the islands of Ireland, clumps of seaweed were combined with the humus in between the gaps of the rocky soil to plant potatoes, a method still used today.[50]

Documents from the sixteenth century attest to the use of composted seaweed as fertilizer in Brittany and Iceland.

In China, there is ancient evidence of kelp ash being spread over the rice paddies.

In Canada, even today, many soils with a pH that is too acidic to grow crops effectively are enriched with extracts of *Ascophyllum nodosum* (also known as *feamainn bhuí*, rockweed, Norwegian kelp, knotted kelp, knotted wrack or egg wrack), which is spread on the fields to make the pH more neutral. Nova Scotia-based Acadian Seaplants, one of the largest seaweed harvesting and drying companies in the Western world since the 1980s, has grown primarily thanks to this activity.

Further south, in the United States, seaweed follows the trend of European crops. There, the progress of algae-based biostimulants has been booming since the recent decriminalization of cannabis in certain states. Indeed, kelp contains exactly what hemp needs.[51] The use of seaweed, in particular *Ascophyllum* and *Laminaria*, therefore makes it possible to boost the natural and organic production of high-quality cannabis.

The example of maerl, or the excesses of industrialization

The use of seaweed in agriculture has sometimes led to serious abuses that we must learn how to prevent. The example of maerl should be mentioned here.

Maerl (*Phymatolithon calcareum* and *Lithothamnion corallioides*) is a shoreline deposit formed of fine gravel and the debris of red algae with a rigid calcareous skeleton. Maerl beds can sometimes spread over several kilometres of coastline. Resembling small corals, the living part grows on the dead part at a very slow rate of around 0.1 millimetres per year. In Europe, maerl beds are mainly found in Brittany, Ireland, Galicia and Norway. In the rest of the world, maerl is very abundant in Brazil, Florida and Chile. It is considered one of the oldest marine plants and one of the slowest-growing.

Its populations are major reservoirs of biodiversity and form the basis of an incredible underwater food network. Containing a very high concentration of calcium carbonate, it is also very rich in magnesium and trace elements. This resource was abundantly 'harvested' and spread on fields in order to increase productivity, in particular for acidic soils. Maerl is also very useful in terms of animal nutrition or for medical treatments. It can even be made into ointments.

Despite its low growth rate limiting the potential for cultivation, a Breton company has started to produce a very specific species of red calcareous seaweed, *Jania rubens*, for cosmetics. The extraction of this calcareous seaweed began over a hundred years ago in Brittany, but was industrialized in the 1970s, using specialized vessels, dredging equipment or 'suction machines' on the seabed. These maerl beds and their biotopes have been irreversibly degraded. In just a few months, these practices destroyed the living parts of the beds, which were several millennia old, and the whole ecosystem that relied on them.

Between 1970 and 1980, several beds in the bays of Saint-Malo and Saint-Brieuc were completely destroyed. These massive extractions were finally banned in France in 2011 and the remaining beds are now protected by European regulations (the Natura 2000 network of protected areas).[52]

This did not prevent these same manufacturers from continuing to harvest maerl from countries with less stringent legislation on the subject, notably Brazil, Chile and Iceland. These dynamic ecosystems remain essential and, in addition to the total disappearance of an environment incredibly rich in biodiversity and ecological services, the removal of maerl has led to the risk of significant erosion of coastal soils.

We are faced with an invisible and silent scandal which, unfortunately, nobody has ever cared about, because everything that lives at the bottom of the ocean seems so removed, so insignificant to us.

In the end, whether it's for humans, animals, plants or soils, the benefits of seaweed are the same. Backed by two billion years of experience and interaction with living organisms, seaweed compounds are beneficial to the vitality, immunity and resistance of all organisms, whether plant or animal.

If we can prevent the excesses of the past, the possibility of cultivating seaweed to support our livestock and agriculture offers the promise of more sustainable food systems.

The promise of animal nutrition using natural and non-synthetic molecules, where food replaces medication.

The promise of an agricultural system operating in a circular fashion between the land and the sea, recycling the nutrients that pass from one to the other.

The promise of meat production reducing its terrifying contribution to climate change.

Above all, it offers the promise of a sustainable and regenerative agricultural system that respects plants, animals and their environment.

Our environment.

Macrocystis pyrifera ('giant kelp' or 'bladder kelp') – this brown seaweed is the largest type of seaweed. It can measure up to 45 metres long,[53] forming underwater forests that grow at a rate of 60 cm per day. Very popular with scuba divers, these forests are home to unparalleled biodiversity, whether marine or terrestrial. It also sequesters a lot of carbon to fuel its very rapid growth. Its pear-shaped, air-filled bladders (*pyri*) have given this seaweed its name. It enjoys the cold waters of western America, southern Australia, New Zealand and South Africa. Giant kelp forests are threatened by human over-exploitation and the resulting ecological imbalances.

3

ECOLOGY: SEAWEED FOR SEQUESTERING CARBON, REVERSING GLOBAL WARMING AND RESTORING THE OCEANS

49 Million Years Ago, the Arctic Sea: The Azolla Event

Seaweed and aquatic plants can influence the climate. Sometimes drastically. Almost fifty million years ago, the proliferation of a modest aquatic fern called *Azolla*, also known as mosquito fern, duckweed fern, water fern, or more poetically, fairy moss, caused an extraordinary drop in temperature leading to the end of a geological era millions of years old.

This little plant thus created the entire climate balance of today, characterized by frozen poles and tropical temperatures at the equator...

No plant has had a greater impact on life than this fern which is just a few centimetres in length. The flora and fauna born out of this climatic upheaval have gradually taken over our planet. We mammals and everything around us have found our place in it. This change, one of the most violent that our planet has ever known, was the work of this fragile and ephemeral fairy moss.

In order to understand this chain of events, we have to go back fifty million years to the Eocene era. At that time, the high presence of carbon and methane in the atmosphere created a very significant greenhouse effect on earth, leading to much higher temperatures than today. Earth was just a huge jungle; the dinosaurs had disappeared fifty million years earlier. Antarctica and northern Alaska were lush temperate forests where tortoises walked beneath the shade of palm trees. Birds dominated the planet, ants began to diversify and the first snakes appeared, as did the great mountain ranges such as the Alps and the Himalayas. At the level of the current Arctic, the continents were joined up around an immense inland sea of four million square kilometres.

The future of the planet and life on earth was to be decided at this very spot.

This Arctic Sea was almost entirely surrounded by land. The high temperature and winds led to significant evaporation at the surface, which generated a high salinity of the remaining water mass on the ocean floor: almost exactly like the Dead Sea today.

The density was exceptionally high; so much so that the fresh water arriving from the surrounding continental rivers could no longer mix with the ocean water column as it usually does.

On the surface of this very salty sea, a thin layer of fresh water therefore developed, called a nepheloid layer. This phenomenon can be found today in the Gulf of Mexico.

Mosquito fern thrived there, in those few centimetres of fresh water on the surface of the Arctic Sea, because this fern is not a seaweed but a freshwater aquatic plant that can only grow in the sea under these special conditions.

The strength of this fern lies in its symbiotic relationship with blue-green algae (cyanobacteria, often referred to as microalgae), which provide it with the nitrogen it needs. The nitrogen available in the air and in the soil is often a limiting factor for the growth of plants, but *Azolla* is exceptional in this regard because, if the circumstances are right, it can create nitrogen completely independently via this symbiosis. This phenomenon allowed this fairy moss to proliferate very rapidly.

The high temperature and the mineral nutrients – in particular phosphorus – arriving in abundance from the continents, created optimal conditions for *Azolla* to double its biomass in two or three days. Soon, the entire Arctic Sea became one giant forest of four million square kilometres that could no longer be distinguished from the surrounding continents. But this forest was not on the land, but on the sea. This would lead to a sudden cooling of the atmosphere.

The fern, like all photosynthetic plants, absorbs carbon from the air and releases oxygen while making material for its stem and leaves. As its life span is only a few months, when it dies it decomposes or is eaten by animals. In one way or another, it therefore releases its carbon into the atmosphere, contributing to the normal cycle of chemical elements and to the balance of our planet.

In the case of the Arctic Sea, the process is different. When it dies, the fern sinks to the bottom of the ocean, where the lack of light and oxygen means that the bacteria or micro-organisms responsible for putrefaction cannot survive. The carbon is therefore trapped at the bottom of the oceans, unable to emerge. It will fossilize before sinking deep into the earth's subsoil.

This is how oil and other fossil fuels we have used so extensively in recent decades were created.

For *Azolla*, the process lasted hundreds of thousands of years in considerable proportions, creating an immense concentration of carbon at the bottom of this salty sea. This carbon therefore moved from the atmosphere to the marine sediments without being able to continue its cycle. The greenhouse effect was then considerably reduced, causing the exact opposite of the climate change we are experiencing today. For the first time in history, the poles gradually became covered with ice. The annual mean temperature fell from 13°C to -9°C between the beginning and end of the event. As the polar ice formed, this in turn trapped a huge amount of carbon and methane, accelerating the cooling process. This glacial shock considerably modified the climate on earth: we went from a 'greenhouse planet' to an 'ice planet'.

The evolution of our ecosystem has been shaped by this difference in temperature between the glacial poles and a tropical equator. This balance gave us a climate with distinct periods of the year, which later came to be called 'seasons'. Terrestrial and aquatic animals evolved, plants adapted, mammals began to thrive, to diversify, giving rise to the fauna and flora that we know today.

Other factors have contributed to this change, and such an abrupt transformation of a planet's climate is of course due to a set of very complex phenomena. This little fern was the catalyst, so much so that geologists named this tipping point the '*Azolla* Event'. There are many lessons to be learned from this event.

Firstly, it should be noted that this tiny innocent fern caused the destruction of an entire ecosystem by drastically altering the temperature of the planet. Let this story be a reminder for all those who still think that despite all the evidence to the contrary from our Anthropocene age, humans are not capable of sufficiently influencing the course of our climate. If a small green plant with a life expectancy of a few weeks could do it, there is a chance that we could too.

Another lesson can be learned by looking at the geographical location of *Azolla* today. After having conquered and entirely covering an area of more than four million square kilometres, polar glaciation and the temperate climate made a huge part of the planet uninhabitable. *Azolla* has now mostly taken refuge in the tropics, struggling to survive on a limited strip of the earth.

A similar fate seems tragically possible in the short term for our species, in view of the latest scientific predictions.

If we try to look at this event through a more positive lens, it demonstrates above all the decisive role of the ocean and aquatic plants in sequestering carbon in a natural and long-term way.

Azolla caused the temperature of the poles to drop more than twenty degrees by decreasing the concentration of carbon dioxide in the atmosphere from 3,500 parts per million (ppm) to 650 ppm. We are now at 450 ppm, and ideally we need to get down to 300 ppm to reach pre-industrial levels. Otherwise, the temperature will continue to rise until the ancient polar ice melts, releasing the vast amounts of carbon dioxide and methane that have been trapped there for 49 million years. This release will trigger an uncontrollable greenhouse effect that will boil the earth back up to an Eocene climate and, in the process, eradicate most of the species alive today.

To avoid this scenario, it is clearly necessary to reduce our greenhouse gas emissions, but also, in parallel, to actively return the carbon accumulated in the air to the earth for many years to come. From this perspective, seaweed has a greater growth capacity than *Azolla*. A large proportion of it also ends up at the bottom of the ocean, trapping its carbon for millions of years.

Seaweed may therefore represent the only large-scale natural solution for capturing carbon from the atmosphere and returning it to the earth's soil, where it has been sequestered for millions of years.

Huge precautions need to be taken, but seaweed and all marine plants do offer a concrete solution for combating what is undoubtedly the greatest threat to the survival of our species.

Seaweed to reverse global warming

Seaweed is among the first victims of climate change and at the same time represents an untapped potential for combating it. Marine plants offer three types of contribution in this fight: decarbonizing the economy, sequestering carbon and restoring ecosystems which will, themselves, absorb carbon.

The first contribution involves using seaweed for applications that reduce our greenhouse gas emissions. This approach is most obvious in the case of 'algo-sourced' products replacing products from fossilized materials.

For example, in the early 2000s but also after the first oil crisis in 1974, seaweed was considered a possible source of biofuel. After all, petroleum is a dead substance made up largely of algae that have sedimented over millions of years through the geological strata.

The benefit to the climate is that it uses material already present in the carbon cycle, rather than contributing to increased levels of particulate matter in the air by digging up the carbon long trapped underground and releasing it into the atmosphere.

Unfortunately, the yields for converting seaweed into biofuel have proved to be too low in the current state of our scientific knowledge, and the economic model is not tenable.

However, as we will explore in the following chapters, seaweed can easily replace other oil-related products such as cotton, plastics or other composite materials which have a significant carbon and environmental footprint.

It also promotes a more local economy, one that is less based on energy-intensive international transport. This is true for both human food and animal feed, as we have seen with the example of Brazilian soya. Finally, seaweed can bring about significant changes by indirectly reducing greenhouse gas emissions in certain industries, as mentioned previously with the elimination of methane emissions from cattle.

We should mention, too, the benefits of biostimulants and the use of seaweed to enrich the soil with nitrogen, phosphorus and other minerals which could replace fertilizers. Or even the case of 'biochar', a technique that involves using seaweed to increase the soil's capacity to absorb carbon and directly improve its fertility.

This point about fertilizers is not an insignificant one. In the case of mineral nitrogen fertilizers alone, it takes just over one tonne of oil equivalent (toe) in the form of natural gas to produce one tonne of nitrogen. This indirect energy expenditure represents 60 to 70% of the energy consumed by field crops, far ahead of other items such as other fertilizers, fuel consumption, phytosanitary treatments and seeds.[54]

In 2015, the French Agency for Ecological Transition, ADEME, noted that reducing mineral fertilization by only 35 kg of nitrogen per hectare on the entire fertilized surface area in mainland France would result in savings of 540,000 kilotonnes of oil, equivalent to the annual production of 200 onshore wind turbines.

We will come back to the importance of seaweed in the nitrogen and phosphorus cycles later.

Another thing to note is seaweed's ability to lessen the impact of waves and thus prevent coastal erosion, thus conserving the productive potential of existing plants on land and their contribution to capturing carbon.

These potential direct or indirect positive effects on greenhouse gas emissions are vital and must be highlighted. Alas, reductions in greenhouse gas emissions and the actual decrease in carbon dioxide in the atmosphere are measured in decades or centuries. So even if we managed to stop all our carbon emissions today, the consequence of this action would only have an effect in the atmosphere in fifty years or more. Which would be too late.

We therefore probably need to retrieve the carbon above our heads and trap it elsewhere…

Seaweed for sequestering carbon and cooling the atmosphere

Carbon sequestration through chemical or mechanical processes seems just to displace the problem and is still highly prohibitive in terms of cost.

Seaweed, on the other hand, is a natural solution and has a development potential that humans have not even begun to explore.

Through photosynthesis, seaweed absorbs carbon to create its biomass. It grows at a very fast rate, more than 50 centimetres per day for *Macrocystis,* or giant kelp, which can reach up to 60 metres in height. Research published in 2022 by a group of leading researchers shows that wild seaweed is present in a surface area of nearly seven million square kilometres, an area comparable in size and productivity to the entire Amazon rainforest, although distributed along the world's coastlines. But seaweed is much better at absorbing carbon than any terrestrial biotope.[55]

The other advantage is that throughout its life, seaweed suffers cell losses, mainly due to swell and currents, which it constantly compensates for, just as we do with our skin cells or hair. The same phenomenon occurs in trees that lose their leaves.

These organic particles lost by the seaweed can represent in total almost 50% of its final biomass. Much of the carbon seaweed absorbs is therefore lost to the oceans.

Around half of this exudate will be used to feed the plankton or the 'filter feeders' (shellfish, bivalves, krill, sponges, etc.), which form the base of the oceanic food pyramid. This half will therefore greatly contribute to restoring the surrounding flora and fauna, which will in turn absorb carbon.

The other half not consumed by the food chain will fall to the bottom of the oceans and sink into the abyssal sediments. If it reaches depths greater than 300 metres, the carbon will be trapped for about a century; if it reaches 1,000 metres, it will be trapped for millennia.[56]

Seaweed therefore plays an essential role in combatting global warming.

Tim Flannery, a well-known Australian environmentalist, suggests that if 9% of the world's oceans were properly managed to produce seaweed, they would absorb more greenhouse gas emissions than we emit today. The oceans would then begin to actively draw down carbon from the atmosphere to cool it.

First, it would be necessary to determine precisely the capacities of seaweed to transfer carbon into marine sediments depending on the species, external conditions, age, location, etc.

Ocean 2050, an NGO led by Alexandra Cousteau, the famous oceanographer's granddaughter, with the help of Carlos Duarte, one of the world's leading researchers in marine ecology, has launched an ambitious research programme to calculate the level of carbon sequestered in twenty-three existing seaweed farms around the world.[57]

Preliminary results released in 2022 indicate a carbon sequestration of 3.5 tonnes per hectare, which is three times more carbon sequestration than one hectare of Amazon rainforest. A figure which could rise, according to the NGO, to 10 tonnes per hectare under optimized conditions. Moreover, this carbon can be sequestered on different time scales depending on the depth of the spaces around the farm. This sequestration could perhaps be improved by cultivating seaweed near major fault lines in the ocean, as long as conditions allow. In this way, these farms would become real, all-natural carbon sinks.

Studies estimate that seaweed could absorb 10 billion tonnes of carbon equivalent per year, or almost a fifth of total annual emissions.[58] In addition, at the end of 2022, the Max Planck Institute for Marine Microbiology in Bremen released fundamental research on the structure of seaweed, which may well represent a potentially revolutionary development in carbon sequestration. Sulphated polysaccharides naturally present in a number of types of seaweed, 'fucans' contain 'focose' sugars that appear to be resistant to degradation by bacteria in any environment. Thus, certain compounds in the seaweed would trap carbon and be able to hold it for many years, in any environment. These very recent discoveries will have to be further investigated and refined, but if they are confirmed, the real capacity of seaweed to sequester carbon could be even higher than expected and become truly significant.

Add to this the potentials for emission reduction, and seaweed would not only be carbon neutral, it would actually be carbon negative. This resource would therefore become the only food capable of reversing the curve of climate change and cooling the atmosphere. Every time we eat seaweed, we are contributing in some way to trapping carbon at the bottom of the oceans!

As a result, some are considering a more radical strategy to cool the atmosphere: growing seaweed for the sole purpose of deliberately sinking it to the bottom of the ocean. This is the *Azolla* strategy!

In this way, these farms would become real, all-natural carbon sinks. According to these researchers, the impact on the ocean would be relatively low, because the ocean's mass is 250 times greater than the atmosphere.

So, if we transferred half of the carbon from the atmosphere to the oceans it would considerably cool the climate, but only modify the carbon content of the oceans by 2%.

This is a form of geoengineering that is very popular in some countries where they even advocate 'seeding iron' into the ocean to accelerate the growth performance of this seaweed.

The solution has numerous advantages, and the production cost is minimal, as no effort is required for harvesting, transporting or drying the seaweed.

The sequestration result is optimal because when we use our seaweed to produce a food or other resources, most of the carbon held in the seaweed is released and re-enters the carbon cycle. So only the biomass that is lost during growth and not captured by the surrounding ecosystems – i.e. a maximum of 10 to 15% – ends up being sequestered in the abyssal zone.

If the seaweed sinks into the deep sea, all of the carbon will be sequestered.

Funding for the implementation of these solutions could be supported by carbon offset credits yet to be created, or by 'impact investors' seeking to fund large-scale projects to combat climate change.

Whatever it costs, this solution is exceptionally effective at returning carbon from the atmosphere back to the oceans from which we extracted it to satisfy our thirst for energy.

But it does raise the ethical problem of voluntarily sinking millions of tonnes of nutrient-rich food while almost a billion people on earth are starving.

The other necessary reservation – perhaps even more fundamental – concerns the balance of the ocean ecosystem. A build-up of carbon on the ocean floor could lead to an explosion in bacterial growth and uncontrolled phytoplankton blooms, resulting in chain reactions that no study can currently measure.

The ocean has numerous and more or less well-known interactions with the surrounding continental shelves, but also with the ocean floor. No one can predict the consequences that this massive engulfment of seaweed in the deep sea would generate.

Moreover, seaweed is not only made up of carbon but also of many other nutrients (nitrogen, phosphorus, minerals, etc.) that are part of a cycle and enable life in the oceans and on land.

Sinking seaweed to the depths of the ocean will result in retaining a large amount of carbon there for thousands of years, but also a large amount of nutrients along with it…

This approach divides scientists: there are those who apply the precautionary principle and those who say that the consequences of climate change are already so advanced that not trying anything would be worse.

It is the eternal debate between geochemists and biochemists. Can the immense complexity of the balance of our biosphere on earth be reduced to particle exchange models?

It is not just the carbon cycle that we are modifying, but the life cycle!

The two are intimately linked, and meddling with the oceans seems like a dangerous idea…

Underwater forests for preserving climate and biodiversity

First of all, our immediate concern is clearly preserving life on earth.

Today, 50% of the carbon on our planet is stored in ocean sediments. The oceans appear to have far greater potential to combat global warming than forests and other terrestrial ecosystems.

The biomass of seaweed in the ocean is greater than that of all the forests on land.

The amount of carbon dioxide captured by wild seaweed worldwide is almost equivalent to the amount of gas emitted by France and the UK combined.[59] But these carbon retention capacities are under threat. The largest types of seaweed have the greatest potential for carbon sequestration. Unlike many plants, long brown seaweed does not thrive in heat or warmth, but in the cold. It grows in southern Chile, Canada, New Zealand and Norway. Wherever it finds cold water. Unfortunately, this water is not as cold as it used to be, and is still getting warmer...

Global warming means that many of these larger types of seaweed risk being replaced by other smaller ones, or even not being replaced at all. This would lead to the disappearance of the entire surrounding biotope, and the cycle of climate change could be further accelerated.

In California, the Kelp Highway has almost disappeared over the past decade. One of the world's largest wild seaweed forests, the Amazon of the oceans, home to a rich ecosystem spanning tens of thousands of kilometres, has lost 90% of its surface area in less than five years. As if it's been wiped off the map.

As early as 1845, Charles Darwin noted that 'The number of living creatures of all Orders whose existence is intimately dependent on kelp is wonderful. I can only compare these great aquatic forests... with terrestrial ones in the intertropical regions. Yet, if in any other country a forest was destroyed, I do not believe so many species of animals would perish as would here, from the destruction of kelp.'[60]

In the American West, 750 species face extinction because they depended directly on these 'miraculous' kelp forests that absorb so much carbon from the atmosphere and protect the coasts from rapid erosion.[61] This worrying disappearance of seaweed is not directly caused by water warming, but is the consequence of the disturbance of ecosystems as a result.

In California, but also in Norway or Japan where the trend is unfortunately the same, the culprits behind this massive deforestation are innocent little red sea urchins. Their destructive proliferation is a consequence of the disappearance of their predators, the sunflower sea star. Previously abundant, they were eradicated by diseases during the intense warming of the Pacific and the two years of extreme drought on the American West Coast, which ended in 2015.

Rarely has such a drastic decline in a species been observed.

In the absence of the sunflower sea stars, the sea urchins proliferated and, as they grazed on seaweed, they devoured practically all of it. In doing so, they destroyed a habitat that provided food and shelter for juvenile fish, octopus, sharks, crabs, shellfish, birds, herons, otters and even sea lions.

Without plants, on land and at sea, there will be nothing left.

Today, California has several thousand kilometres of what some call 'the lost desert of sea urchins'. Ironically, these sea urchins are not even marketable because there are too many of them and they don't eat enough to reach a sufficient size. Sea urchins are unique animals: they can sometimes live for up to 200 years; their vital and reproductive functions do not age; and they are able to modify their feeding behaviour to overcome crises. When resources are scarce, they have the ability to lie dormant for many years, remaining puny and subsisting on any waste matter in sight.

Tasmania used to have huge reserves of giant seaweed, but in similar circumstances caused by global warming and the overfishing of lobsters – which are also fond of sea urchins – it has now lost almost 97% of its *Macrocystis*.

Sometimes climate change is not the only culprit. Trawling and dredging of the seabed can quickly wipe out a kelp forest. Greenpeace therefore decided to install 180 blocks of granite off the coast of Sweden in order to protect certain species of seaweed by preventing these machines from accessing 'marine protected areas' (MPA) where fishing rules were too rarely respected. The destruction of ocean ecosystems is a survival issue for all of us, one that goes far beyond climate change.

Yet, although we are rightly alarmed by a fire on land in the Amazon or Australia, it is rare to see front-page news stories covering the disappearance of aquatic forests, which are the second lung of our planet; because we know so little about these forests and we see hardly any images of them. But the forests are also burning under the sea. Far below the waves, life gradually disappears to give way to a liquid desert of sand and rock, where nothing grows and wildlife is disappearing.

The situation is not hopeless, however, and it is certainly not time to give up. Seaweed has been able to withstand many changes over these billions of years.

Somewhat paradoxically, the predicted melting of the sea ice could lead to a cooling of the oceans, giving hope for a return to more favourable water temperatures and also freeing up space for the development of our underwater forests. Thus, the types of seaweed disappearing from certain geographical locations will be replaced by others and will be able to develop in new zones closer to the poles, where the conditions are more favourable for them.

And even in areas where kelp forests are disappearing, there are solutions. The great advantage of seaweed is the speed with which it can grow. When a terrestrial forest disappears, trees can be replanted, but it will take years or even decades for it to return to its original state. In the case of a kelp forest, three months can be enough to re-establish the existing vegetation and allow life in all its forms to flourish again.

There are many examples. In the 1970s, Norway lost 8,400 square kilometres of its *Laminaria hyperborea* forests due to an explosion in sea urchin populations.[62] Climate change plus the development of predatory crabs have enabled 3,400 square kilometres to be restored in just a few years. The so-called 'green gravel' method of seeding gravel and dropping it into the sea has proven to be very effective. However, the 15,000 kilometres of coastline north of Tromsø still suffer from the proliferation of sea urchins. Up to thirty sea urchins can be counted per square metre, when the normal concentration is two. We still need to find intelligent ways of dealing with the problem of sea urchins. Urchinomics, a young Norwegian company working in California and Japan, is making great strides in this area. Their idea is simple: based on the principle that a well-fed sea urchin is an expensive delicacy in Japan and in a growing number of other countries, the company's idea is to take sea urchins from the ocean and feed them with seaweed waste in land-based ponds. The company has developed a very specific aquaculture methodology for feeding them and making them grow quickly. Once the sea urchins are fully grown, they are sold at a premium, which pays for the divers to go and collect more. The start-up is therefore transforming a pest into a delicacy. It has even gone a step further, because with sea urchins, only the 'gonads' (the inner part) is edible. Urchinomics has therefore managed to find a way to grind up the spines and use them as a very effective fertilizer to accelerate the growth of kelp.

This methodology has yet to be refined and large-scale pilot projects have yet to be developed.

But these experiments prove that it is possible to help 'repair' anthropogenic damage quickly. Above all, it shows that it is possible for largely degraded marine ecosystems to regain their equilibrium, but that a human touch, carried out with great scientific rigour, is often necessary to enable this reversal.

In the light of these recent experiences, the idea that 'leaving the oceans alone' is the best way to restore their lost vitality is proving to be a dangerous gamble. The impact of humans on the oceans is already too great to leave the latter to fend for themselves...

We still need to work out how to support our oceans in order to live in harmony with them. We urgently need an approach aimed at protecting, restoring and maintaining these forests, to enable them to carry on impacting our climate.

'Blue carbon': using the ocean to offset our emissions

Seaweeds are obviously not the only organisms in the ocean to carry out photosynthesis and to retain carbon.

Microalgae have exceptional potential for carbon capture and can offer solutions on land, but cultivating them remains almost impossible in an open environment given our current state of knowledge.

Aquatic plants such as ferns and water lilies absorb a lot of carbon but have limitations linked to the availability of fresh water on the planet. Lack of land space, rapid population growth and droughts are already threatening our freshwater resources. This makes it hard to consider using a large portion of them for carbon sinks.

Coastal ecosystems such as mangroves, brackish marshes and seagrass beds (*Posidonia*, *Salicornia*, etc.) are exceptional candidates for both sequestering carbon and delivering important services to the ecosystem. Mangroves store three to five times more carbon per hectare than terrestrial forests and provide essential breeding grounds for marine wildlife.

Restoring and preserving these ecosystems is critical for the implementation of an integrated solution to combat both global warming and the loss of biodiversity. Unfortunately, the mangrove forests are losing 2 to 3% of their surface area every year: it has shrunk by 30 to 50% between 1980 and 2000.

The areas suitable for these species are also limited to very small coastal strips which have very specific conditions, and which have already been exploited for other human activities such as fishing, maritime transport, aquaculture or tourist activities.

However, it is important to note that, unlike seaweed, these coastal ecosystems already benefit from offsetting schemes recognized by the United Nations. Their restoration can thus be financed by 'carbon credit' schemes, created under the Kyoto Protocol.

Thanks to this agreement, organizations that emit greenhouse gases can finance the planting of mangroves or other forests in order to 'offset' their current emissions and meet their commitments not to contribute to global warming.

These commitments are very popular and are far from trivial, gradually becoming a necessary prerequisite both for investors and clients of these organizations. This scheme therefore allows financial support to be allocated to replanting trees, mangroves or seagrass beds.

Unfortunately, seaweed is not yet eligible for these offsetting systems, due to a lack of sufficient research and existing international standards. As mentioned earlier, although numerous studies show that the carbon sequestration by the massive forests of brown seaweed is more efficient than the tropical forests on earth,[63] the seaweed residues are scattered very far, making it more complex to scientifically assess seaweed's true contribution when it comes to retaining carbon. However, the major programmes underway and the initial results presented during the recent COP climate conferences should give the topic a boost in the coming years.

At COP 27 in Egypt, a scientific publication was released demonstrating that one specific compound in seaweed is particularly hard for other ocean inhabitants and bacteria to break down,[64] thus the carbon sequestration is higher than expected. The research reveals that the seaweed mucus called fucoidan is particularly responsible for this carbon removal and estimates that types of brown seaweed could remove up to 550 million tonnes of carbon dioxide from the air every year. This amount is almost equivalent to Germany's total annual greenhouse gas emissions!

In addition, in late 2022, Japan announced the world's first voluntary blue carbon credit for kelp-bed restoration by a company specializing in removing urchins to regenerate seaweed forests.[65] These first blue carbon credits certified by local companies averaged a sales price of 536 dollars per tonne while carbon prices in other traditional voluntary carbon markets usually do not exceed 120 dollars. It shows that blue carbon is seen as a higher-quality carbon credit, as ocean-based ecosystems like seaweed forests deliver a wide range of ecosystem services beyond carbon sequestration. These future financial instruments are critical because they will support the initial investments needed for research and will also make it possible for seaweed producers to survive by collecting revenue until they have sufficient production capacities to make a profit.

This lack of funding to reward the services provided by seaweed to combat global warming is all the more regrettable as offsetting schemes are beginning to be sorely lacking. Land available for large-scale tree planting is becoming increasingly scarce and large carbon-emitting companies are now struggling to find projects to achieve the carbon neutrality demanded by their stakeholders.

We need to consider all the ways of planting and cultivating seaweed in new areas with the – at least partial – aim of capturing the carbon and sending it to the ocean floor, just as *Azolla* did 49 million years ago.

The remaining biomass will have to be used to boost the surrounding ecosystems, which will themselves absorb carbon and feed the planet. However, no individual solution seems sufficient. It seems our salvation must come from a combination of solutions implemented at the same time. The regeneration of seaweed forests has a role to play, but we can take it further. We need to preserve them where they currently exist, and also plant and grow them in new areas.

Above all, we need to develop, upstream, the research needed to assess the scenarios linked to this massive unloading of carbon into the ocean over a short period of time, and to develop the tools to monitor and mitigate impacts.

There is no room for error. This stage is unquestionably the most complex and the most ambitious. And the most dangerous.

But do we have a choice?

To sustain our energy-intensive development, we have spent the last seventy years extracting carbon buried for millennia in the earth's sediments.

For our species and those around us to survive for a while longer, seaweed can help us to manage this carbon cycle and ensure that it gets back to where it was.

Under the sea.

Laminaria digitata (also known as 'oarweed', 'tangle-weed', 'sea tangle', 'sea girdle', 'sea ribbon' and 'red ware') – a large brown seaweed that can measure up to four metres and has a plastic-like consistency. Its Latin name comes from its shape, which resembles a hand with long fingers. Especially abundant in the North Atlantic, it is rich in iodine, fibre and vitamins and is often used in soups or for simmering dishes because its flavour is rich in umami. Studied in depth for its iodine compounds and its ability to emit iodine into the air, it is the seaweed that is richest in potassium and was therefore widely used to extract potash. Today, it is an essential resource for the extraction of alginate.

4

SEAWEED FOR HEALING PEOPLE

1850, from Seaweed to the 'Alpine Cretin'

In the nineteenth century, intellectuals, bourgeoisie and aristocrats set out to discover the Alps. The eternal snows on the summit of the peaks inspired writers and painters. There was a blot on this idyllic landscape, however: the large number of 'cretins' found there.

Before it became an insult, this outdated term designated the victim of a terrible disease. The term 'cretin' appears to come from the word 'Christian', perhaps because they were innocent and blessed, or because such patients were often taken in by monasteries.

At the time, France had around 20,000 'cretins' and 100,000 goitre sufferers in its mountainous regions, particularly around the Alps, Pyrenees, Vosges, Jura and Massif Central. Given the low population density in these regions, these are significant numbers. The same phenomenon was observed in the Rocky Mountains, the Black Forest, the Andes, the Himalayas and the Urals.

What was referred to by the now offensive term 'cretinism' is an extremely serious condition caused by the degeneration of the thyroid gland, which deforms the base of the skull and stunts both physical and mental growth. Those who suffer from it are characterized by large goitres and dark, wrinkled, expressionless faces. Those affected are usually less than a metre tall, have deformities of the bones and spine, walk crookedly and suffer from very significant mental disorders. Sufferers of the condition die young...

The disease has been known since ancient times, but was not really identified until the beginning of the seventeenth century.

The rise of interest in the Alps, the fascination for the monstrous and the failures of medicine to find a cure for it gave so-called 'cretinism' great visibility during the eighteenth and nineteenth centuries. This passion for 'monsters' is also found at the heart of the popular imagination and literary productions of the time. These innocent people with deformities therefore became an attraction, the kind of creatures one tries to spot along the roadside, like a bear in the depths of the forest.

A fear of contagion then developed.

In some villages, more than a third of the children were affected by this disease. Could this have been the beginnings of a global pandemic or widespread degeneration? Unless it was a consequence of inbreeding, common in those remote areas, of poor hygiene or of alcoholism. Treatments remained ineffective, and the number of sufferers from this disease continued to increase in the high-altitude villages.

The link between goitre and a lack of thyroid activity was established quite early on, but the origins of this disorder were still unknown.

The cause of this disease could not be established by European medicine at the time, which was arrogantly colonizing the world both politically and scientifically.

Yet the mystery had already been almost solved by Chinese medicine thousands of years earlier!

It was not until the very beginning of the twentieth century that the miraculous remedy was found. It was, however, not all that sophisticated: marine iodine.

The disease, now called congenital iodine deficiency syndrome, is in fact a degeneration caused by an iodine deficiency that prevents the thyroid from functioning properly. Iodine is vital for the human body. It is one of the rarest micronutrients on earth, but is abundant in seawater. It accumulates in fish, marine crustaceans and especially in seaweed. One teaspoon of certain types of dried kelp contains about sixty times the daily dose required by an adult man (150 mg).[66] No living organism on our planet has a higher iodine content than this kelp which washes up on our beaches.

Iodine's name derives from the Greek word *iodes*, meaning 'violet-coloured'. In 1811, Bernard Courtois, a French chemist, burned brown seaweed, washed the ashes with water and added sulphuric acid to remove any contaminants. This precipitate caused a cloud of purple vapour. Courtois had just isolated iodine, and this would prove to be a major discovery.

Iodine is highly volatile and accumulates in the atmosphere over the sea. It then contributes to forming clouds and ends up in rainwater, which transports it to the seas and sometimes to the soil. Plants and meat generally contain a residual part, even in landlocked regions, thanks to rain or the previous presence of glaciers that melted over 12,000 years ago.

In the mountains, geology, torrential erosion, soil wash-off and steep slopes prevent the iodine from staying put. In the 1800s there was no food from the sea there, and mountain agriculture did not provide the human body with enough iodine to function normally. The limestone massifs are less affected than the granite massifs, which explains the disparities within these mountain areas that have so baffled researchers.

This iodine deficiency had been studied from the beginning of the nineteenth century by Swiss physicians. Unfortunately, the treatment was badly applied, in doses that were far too high, and the hypothesis was discredited. Too much iodine has equally devastating effects. The Swiss doctors were not listened to by the international community and it was not until a century later that an adequate treatment was developed, using iodine in very small quantities.

It was not until the beginning of the twentieth century that iodized salt would be found on people's tables. Iodine was then extracted from seaweed ashes and distilled into salt in small doses. It may seem surprising that salt, which itself comes from the sea, needs to be fortified with iodine. In fact, iodine evaporates during the preparation of sea salt. Some must therefore be added after processing to reach the recommended daily amount. Iodized salt contains about 1,000 times more iodine than non-iodized sea salt.[67] At the same time, lozenges were specially prepared for children. Within a few months, the method eradicated this awful disease in France and Switzerland. Nature therefore cruelly reminded us of our original connection to the ocean.

At the start of the twentieth century, the word 'cretin' entered political vocabulary. Marx, Engels and later Trotsky associated 'cretinism' with the criticism of progress, of man's submission to machines, electricity and speed. In the West, *Tintin*'s author Hergé is suspected of having contributed to the popularization of the expression through the character of Tintin's best friend, Captain Haddock (a seafarer) who calls

Professor Calculus (who is from the Swiss mountains) '*crétin des Alpes*' ('Alpine cretin') in the original French-language version of *The Seven Crystal Balls*.

Iodine had many uses then, but today only 0.5% of iodine production is used in food. With the multiplication of needs in the twentieth century, resources have diversified. Drying and burning marine plants to extract iodine was too artisanal. Instead of seaweed, we started using Chilean mines, or nitrate beds, which were formed when the ancient sea was mineralized thousands of years ago. Iodine is extracted, as nitrate, from the raw ore, which comes from ancient groundwater overloaded with these compounds.

More recently, we've seen the development in Japan and the United States of 'brine' systems. In some places, the subterranean crevices of our planet formed by the meeting of oceanic and continental plates of the earth's crust have retained seawater that partly evaporated a long time ago but left behind aqueous solutions highly concentrated in sodium chloride and rich in iodide. The latter compound can be extracted by drilling into the salt deposit.

Thanks to these two processes, iodine is easier to obtain in industrial quantities than it is from seaweed, despite a questionable ecological impact and the fairly obvious non-renewal of resources.

Iodine production is abundant today. However, congenital iodine deficiency syndrome has not disappeared, not by a long way.

It is estimated that it critically affects two million people worldwide, often in mountainous regions far from the sea.

This is not the only possible reason, as cases have recently been found in areas washed out by rain and floods.

In addition, a milder iodine deficiency causes depression, fatigue, constipation, weight gain, a feeling of being cold, and sometimes severe effects on menstrual cycles and cholesterol. It can also lead to serious mental retardation in the unborn child. Nearly two billion people in the world, including a third of children, have an iodine deficiency which causes these types of symptoms, often without knowing it. Similar reactions are observed in farm animals that are gaining less weight and reproducing more slowly.

Iodine also protects the thyroid from being affected by endocrine disruptors. These molecules from our environment (food, indoor air, cosmetics, hygiene products) modify hormonal communication and weaken the thyroid. This essential trace element has the potential to reduce the harmful effects of these products on our health.

These properties explain why, in the event of a nuclear accident, iodine tablets are distributed in order to saturate our thyroid and prevent radioactive iodine from being incorporated into the thyroid hormones and thus causing cancer, especially in children.

According to the WHO, fifty-four countries in the world have populations that do not consume sufficient iodine. Fourteen of these are countries on the African continent, where, according to the FAO, 86 million people, or more than 13% of the total population, are goitre sufferers. Other countries, such as the UK, Australia and New Zealand, and even the Faroe Islands, are noticing a resurgence of this consumption deficit in their countries despite their island status. The standardization of eating habits and the level of food processing are undoubtedly factors in this worrying situation.

In 1995, the Chinese government launched a major iodized-salt programme aimed at eliminating iodine deficiency disorders.

In Africa, the WHO has made combating iodine deficiencies a priority for improving children's cognitive abilities and growth.[68]

The solution seems simple but, as is often the case, its implementation is much more complex. The main challenges in emerging countries are ensuring the long-term viability of salt iodization infrastructures, informing people of the benefits of this practice and providing iodized salt to all local communities, including those who are poor, displaced or living in remote areas.

Although overlooked and misunderstood, seaweed holds many solutions to the problems of malnutrition. These examples of problems related to iodine deficiency demonstrate that we humans have too often neglected the essentials in the development of our medicine.

Life on earth originated in the sea.

Seaweed: a natural medicine

'Let thy food be thy medicine and medicine be thy food,' Hippocrates is renowned to have said. Even in ancient times, the idea was not new. How did we go so rapidly – in less than 200,000 years – from a small population of common mammals in the middle of the food chain, to hyper-predators dominating a planet we have overpopulated?

Human beings have developed, among other talents, the ability to identify the nutrients that allow us to increase our physiological and psychological performance while optimizing our reproductive potential. The fact that humans are among the only omnivorous animals is surely no coincidence when it comes to our meteoric evolutionary success. And seaweed has long been a proud ally in this process!

The way seaweeds were chewed and not ingested in the Monte Verde cave proves that they were also used for their therapeutic properties 14,000 years ago, although it is difficult to know their detailed applications at the time.

The millennia of experience of Chinese medicine tell us much more about the benefits of these organisms, which have become one of the pillars of health in Asia.

One only has to walk into a traditional pharmacy in Beijing to see the role that sea vegetables play in the health of the world's most populous country. The Chinese are very advanced when it comes to alternative medicine.

Our scientists, who are sometimes quick to dismiss these treatments as old-fashioned remedies, are nevertheless right to argue that we lack publications and robust research into the effects of seaweed compounds on our health. However, there are surely lessons to be learned from four thousand years of experience in marine herbal medicine in Asia.

For almost two centuries, our Western pharmacopoeia has turned massively to chemistry, too often neglecting biology. It has considered humans and our health as an agglomeration of responses to independent bioactive compounds. However, the complexity of nature goes much further. We are increasingly realizing this to be true, in Western medicine. The same is true for seaweed, these organisms are only truly effective when consumed whole, as the result of complex and complementary phenomena between different compounds. Studying these compounds individually means losing the complexity of these synergies. In short, it is better to eat whole seaweed rather than try to extract the interesting substances one by one to eat them as food supplements.

This argument is often put forward to explain why our medical research has been so slow to fully identify the therapeutic benefits of the various elements in seaweed. But the lack of interest generally shown by our pharmacopoeia in marine plants is undoubtedly also an equally valid explanation.

The potential is immense, however, as seaweed is the oldest – and therefore most experienced – organism to survive on our planet. For over a billion years it has developed particularly sophisticated defence systems against a wide range of enemies such as grazers, fungi, bacteria and viruses.

These defence systems and the active compounds involved in them are as numerous as they are unknown. However, it is already recognized and accepted that seaweed has anticancer, anti-inflammatory, antiviral, analgesic, immunomodulatory, antibacterial and antifungal properties.[69]

In a world still stunned by the outbreak of a pandemic that has shaken its very foundations, it seems timely for us to consider this complex environment: the cradle of life on earth!

The health benefits of seaweed compounds

In addition to their nutritional value, sea vegetables have widely recognized health benefits.

The most studied component for more than 200 years is undoubtedly iodine. This trace element, as seen above, enables our thyroid to function properly and to control essential functions such as hormone production, repair functions, cell renewal and energy production. Long before it revealed its secrets for our thyroids, iodine first revealed its antiseptic properties. From its discovery in the early nineteenth century, tincture of iodine was the only disinfectant used for wounds and burns. Within a few years, it invaded the world's battlefields, and also pharmacies, medical practices and schools. It is still used today, despite the widespread use of antibiotics and other disinfectants. Its use has been extended to many other areas: X-ray examinations, scans and other cancer treatments.

However, in addition to iodine, there are many other – still largely unknown – components of seaweed that can contribute to our health. Seaweeds and their compounds could play a role in preventing the major causes of death in developed countries, namely obesity, diabetes, heart disease and cancer. For example, the undigested fibres remain in the intestine for a long time, promoting a feeling of satiety that could have a positive effect on obesity issues.

In addition, the high presence of fucoxanthin, the pigment that gives some brown seaweed such as *kombu* their colour, has been shown to have a significant effect on fat metabolism and blood sugar levels. This pigment also shows very promising effects in reducing the development of malignant tumours.[70]

The presence of alginate, which forms the structure around all the cells of a brown alga, prevents sudden spikes in blood-sugar levels. At the same time, certain peptides, protein fragments, could block the mechanisms that raise blood pressure. In addition, certain complex sugars present in brown seaweed, such as fucans, limit blood-clotting problems.

Finally, it turns out that sea vegetables are precious allies in the fight against cholesterol. *Wakame* and *nori*, the most consumed seaweeds in Japan and Korea, contain a large amount of phytosterols. These compounds are chemically similar to 'bad' cholesterol and prevent the absorption of cholesterol into the body. Phytosterols have also been shown to have a significant ability to reduce the risk of breast, ovarian, lung and stomach cancer.[71]

While one in eight women in the US is diagnosed with breast cancer in their lifetime, in Japan it is one in thirty-eight.[72] Most interestingly, this difference is almost non-existent for Japanese women who live in the West and have adopted a local diet.

These properties of sea vegetables could also partially explain the very low incidence of obesity, diabetes and cardiovascular problems in Japan and South Korea.

While the obesity rate is 38% in the United States, 33% in Mexico and 15% in France, it is only 5% in South Korea and 4% in Japan.[73]

These countries also have one of the highest life expectancies in the world and the greatest concentration of 'blue zones' (regions with the highest proportions of centenarians).[74]

According to the WHO, the average life expectancy for men is 84.3 years in Japan, 82.5 years in France and 78.5 years in the United States.[75]

Obviously, we must be very cautious here, as these differences are always based on multiple factors, and the role of seaweed in these figures is difficult to quantify.

To try to be comprehensive, if we really want to consider the potential of seaweed compounds for our health, we must realize that our health goes beyond our diet. The antiviral potential of certain seaweed compounds should not be overlooked at a time when the word 'virus' has become part our daily lives. Sulphated polysaccharides, including carrageenans from certain red seaweeds, have also been shown to have protective effects by reducing the ability of the virus to enter cells and by strengthening our immune system.[76]

A study published at the end of 2021 also shows that one of these polysaccharides from a very common green seaweed (*Ulva sp.*) has shown significant antiviral activity against Covid-19. In-depth in vivo studies still need to be conducted and could give hope for an innovative solution to complement current vaccines and other treatments.[77]

Iota-type carrageenans are complex sulphated molecules capable of forming soft gels. These gels offer proven – for the moment partial – protection, against herpes, papillomavirus or HIV.

Laboratories are currently working on its use in nasal spray form with regard to the transmission of Covid-19, with interesting preliminary results.[78] In a spray, the carrageenan molecules are transformed into an 'antiviral biofilm' that creates a physical barrier on the surface of the nasal and oral mucous membranes and at the same time reduces the activity of the virus. These mucous membranes are the gateway to infections and the first line of defence for our immune system. This type of carrageenan has already shown in vitro antiviral activity against various types of respiratory infections, including human rhinoviruses and coronaviruses.

In early 2021, a large study was conducted on more than 350 hospital staff in contact with Covid-19 in ten hospitals in Argentina, at a rate of four sprays per day. The authors claim that the process confers an 80% reduction in the risk of contracting the disease, and probably of transmitting it.[79]

Further studies are underway in Europe to confirm this promising solution, which would be much less tiresome and polluting than wearing masks.

Again, the relatively low incidence rates of Covid-19 in some Asian countries may suggest that microbiota and dietary habits are factors in the contagion. It would nevertheless be unwise to venture into this field in the current state of our knowledge.

In another field, nearly 2,000 years ago, the Romans used seaweed for the treatment of burns, wounds and skin problems. They had already unknowingly discovered some of the miraculous properties of alginate.

It is a polymer (also a polysaccharide) which makes up the flexible structure of brown seaweed and which was identified in Scotland as early as 1860.

For a long time, alginate from kelp or wrack has been used in dressings because it activates the cells involved in wound healing, maintains a moist environment with great absorption power, promotes coagulation using calcium, and reduces the risk of infection through fixation of bacteria. In 2022, Brothier, a French laboratory specializing in the transformation of brown seaweed into compresses, announced an investment of 10 million euros into setting up a non-woven textile process that will lead to the 'future compress' integrating cell therapy.

This polysaccharide is also used to make dental impressions and to regenerate cell tissue. Currently, animal testing is being carried out to make artificial blood vessels. The most common medications to protect the stomach lining or prevent reflux are also alginate-based. The capsules we ingest are made of alginate, which disintegrates in order to release the active ingredient of the drug into our stomach.

On a different note, a Norwegian company recently developed revolutionary alginate-based solutions to treat cystic fibrosis and other respiratory diseases. Clinical trials are in the final phase and this remedy should soon be released on the market.

The proposed solution is as follows: the bacteria that aggravate the symptoms of cystic fibrosis usually settle in the lungs, developing a biofilm which clogs them and resists treatment. This biofilm, which protects the bacteria, is composed mainly of an alginate secreted by the bacteria, that is very similar to a certain type of alginate from brown seaweed. Thus, by spraying fragments of alginate into the lungs before the bacteria develop, the bacteria no longer have the capacity to attach themselves to the mucous membranes, create their own biofilm and multiply. Patients can breathe and the antibiotics are able to take effect because the bacteria are no longer protected.

If it is definitively authorized, this treatment will be a major advance in respiratory and pulmonary medicine.

Brown seaweeds and their alginates are not the only ones that are good for our health. In Australia, significant funding has just been allocated to research centres to understand how seaweed can help heal wounds, prevent ageing and regenerate human tissue.

Seaweed modulates our gut microbiota

The benefits of seaweed for our intestines have long been recognized throughout the world without having been, until very recently, properly explained.

Various well-known laxative brands contain extracts of *Fucus*, a genus of brown algae also known as rockweed, which is widespread throughout the world.

Certain amino acids present in red seaweeds have been

used as deworming agents for a very long time throughout the world, whether in Turkey, Norway, India, Africa or
China. Ever since ancient times, writings also report the use
of Corsican moss (*Alsidium helminthochorton*), harvested
around Ajaccio, for treating worms and parasites.

In addition to facilitating intestinal transit, algae can act
as modulators on the ultra-complex system which constitutes the most promising field of medical exploration at the
beginning of this twenty-first century, the one we are already
calling our second brain: our gut microbiota.

The microbes in our gut (two to three kilograms of them)
are mostly acquired at birth and in our early years. They are
a key asset that interacts permanently with our bodies and
our brains.

The microbiota consists of viruses, parasites, fungi and
bacteria. The last are important and very numerous: it is
estimated that we have more bacteria in our intestine than
human cells in our body. There are such astounding interactions among these organisms living in symbiosis that we could
almost define ourselves as a bacterial being simply encased
in a human exoskeleton...

More seriously, the DNA sequencing capabilities developed
over the past few years, coupled with the enormous data-
processing capacities of our supercomputers, have opened the
door to a new field of knowledge in the biological complexity
that enables life.

And an essential part of that is found in the contents of
our intestines.

It took thirteen years to decipher the three million pairs
of nucleotides that make up our DNA. The first human
sequencing was finalized in 2003 and cost over three billion

dollars. Less than twenty years later, it is possible to sequence an entire DNA in minutes, for the cost of just a few dollars. This change of scale offers the possibility of going much further in understanding the bacterial world that defines part of who we are.

When it comes to the link between the contents of our intestines and our health, here again, the Chinese were pioneers: they were the first to use faecal transplants to treat diarrhoea and food poisoning, as early as the fourth century. This effective treatment was called 'yellow soup' and involved transferring microbiota from a healthy person to an unwell one.

There is no need to dwell on the practical details of how this transfer took place…

Inside our intestines is an unstable equilibrium that can be transferred and maintained.

On this topic, it was long thought that the appendix was a vestigial and useless organ. This assumption is false according to the work of Midwestern University,[80] and the appendix may in fact have been vital to the survival of our species, acting as a bacterial 'reservoir' enabling us to preserve some of our microbial heritage in our digestive system, even in cases of violent dysentery.

So even though diarrhoea 'cleanses' the inside of our guts, some of the precious microbiota remains squirrelled away in the appendix. Once the colic is over, the colony of microbes emerge from the appendix and slowly spread through the intestine again.

Those who have already had their appendix removed can rest assured that diarrhoea is rarely so violent these days, plus our knowledge of probiotic treatment has progressed considerably.

Every day brings its share of surprises, as we progress in our understanding of this 'second brain' and its complex interactions with our actual brain.

The main challenge is to influence the content of this microbiota in order to maintain or improve our health. This microbiota of 'good' and 'bad' bacteria loves the fibres and other molecules found in seaweed. For example, sulphated polysaccharides have demonstrated properties that greatly enhance the growth of the 'good' bacteria that line our gut and make it possible to treat a wide range of diseases.

Nantes University Hospital is currently looking at how other compounds in seaweed might modulate our microbiota and act on our resistance to stress. If we are anxious, our immune functions are affected, and this accelerates the development of chronic diseases. Influencing the microbiota to reduce stress means helping the sick to get well and the healthy to stay well. This phenomenon has already been observed empirically, for centuries, in plants via biostimulants and in livestock via animal feed.

This powerful connection between seaweed and other living organisms reinforces the concept of One Health developed in the early 2000s in response to medical hyper-specialization and the fragmentation of human and veterinary health disciplines.

The approach promotes an integrated, systemic and unified vision of humans within their animal and plant environment. The development of viruses that originate in animals (including Covid-19) is a powerful indicator of the relevance of this vision.

Some medical specialists even suggest that the remarkable influence of seaweed on our intestines can be explained by the fact that these marine plants predate all other life forms in history. All the organisms that developed after seaweed would therefore have learned to evolve in interaction with them. The latter would be a kind of matrix for all the different life forms.

The immense genetic diversity of seaweed provides an almost infinite field of exploration in order to search for new modulators that can influence our gut and our health. More prosaically, the other advantage for our microbiota is that the fibres in seaweed are generally also resistant to the digestive mechanisms of our stomach and so become a food source for the microbial community in our gut.

Seaweed is a fantastic prebiotic.

Again, lactic acid fermentation techniques can accelerate these processes and create the enzymes needed to make the metabolites and proteins in seaweed more available to our body. Fermentation also produces certain types of amino acids that are more important for our bodies. There are numerous examples of this approach being applied, and the results are beginning to be published. They could soon become part of our medical landscape.

In China, for example, a new drug using an active ingredient based on fragments of a polysaccharide from brown seaweed, sodium oligomannate, has just been marketed to combat Alzheimer's disease, with impressive results. The project originated from the observation that regular consumers of brown seaweed had a lower prevalence of Alzheimer's disease. Research has identified this oligomannate as a potential active ingredient, explaining a greater resilience of cognitive abilities in these kelp eaters.

However, these compounds will not directly influence the brain. Basically, seaweed will regulate our microbiota, which will reduce the kinds of inflammation in our brain that promote the development of a number of degenerative diseases, including Alzheimer's disease.

The first months of experimentation show very positive results, without any side effects. The drug is currently in the validation phase in the United States and Europe, where it has caused much debate and hesitation, no doubt more on geopolitical grounds than medical...

Much remains to be discovered about the potential medical applications of seaweed compounds, and even more so about the possibilities linked to using the bacteria that live in symbiosis with them.

Seaweed in cosmetic and hygiene products

The list of benefits of seaweed for our body is still far from exhaustive.

Ulvan – a polysaccharide extracted from green seaweed – could develop biological structures used in skin grafts and dressings. This compound acts at the cellular level, accelerating the healing of wounds and preventing the formation of scars. However, it is not yet available in the same quantities as alginates or carrageenans.

The gelling agents of red seaweed, agar or carrageenan, are used in the manufacture of suppositories. The Colgate brand built much of its wealth and reputation thanks to its efforts to incorporate extracts of red seaweed into its toothpastes in order to make the paste less liquid: a process which later became widespread.

Every day, when your brush your teeth, the toothpaste does not flow down the sink thanks to the carrageenans in seaweed.

Still on the subject of teeth, the bacteria living on the surface of seaweed have also been identified as being very efficient at secreting enzymes that attack dental plaque.

Seaweed is the new star of the field of cosmetics. Creams based on seaweed with moisturizing, regenerating and, soon, anti-UV properties represent by far the greatest added value of the sector and contribute to its development. The active ingredients are still far from being fully identified, however. The Canadian *Chondrus crispus*, also known as Irish moss or carrageen moss, has certain molecules that resist the ice-cold waters of winter. The L'Oréal brand uses these molecules to protect dry skin from the aggressions of the cold.

In some respects, the reaction of the tissues of certain seaweed is surprisingly similar to that of our skin. It has even been determined that a green seaweed turns brown in the sun by secreting a kind of melanin that is almost similar to ours. So even seaweed likes to get a tan...

The University of Stirling in Scotland has demonstrated the benefits of the fatty acids of certain red seaweeds, in particular *Asparagopsis*, in combatting strains of bacteria associated with acne. Many exfoliants used for scrubs use seaweed, and thalassotherapy using alginate masks and seaweed wraps is becoming very popular.

Our microbiota is not only essential in the gut, but also in the functioning of the dermis or the genito-urinary system. There are more impressive discoveries to come regarding the interactions with these marine plants, with which we have so much in common.

Seaweed as medical tools

As early as the eighteenth century in Scotland, doctors had noted the absorptive capacity of seaweed and used the stipes (the stems of seaweed), which they dried, to drain abscesses in the abdominal wall.

Seaweed also played an important role in gynaecology for Europe's coastal populations. There are many accounts describing the use of kelp-based tools to induce abortion. It was common to insert laminaria into the cervix to relieve painful menstruation.

Even more commonly, the stipes of *Laminaria hyperborea* were inserted into the vagina for cervical ripening before childbirth. This dried brown seaweed absorbed secretions, swelled and dilated the cervix, inducing labour. This technique has survived, and this material is still used in some circumstances (laminaria sticks). Today's cervical dilators are factory-made but still based on those kelp-based tools.

Finally, agar-agar, obtained from some species of red algae, has gelling properties that are highly valued in food, especially as an alternative to gelatine for vegetarians. It is an essential compound in most microbiology equipment, in particular the agar plates of Petri dishes, those famous little cylindrical, transparent, shallow dishes found in all laboratories. Its inert, non-nutritive substance provides an ideal surface for the growth of bacteria.

Agarose, a purified extract of agar, provides a high-precision gel that is used in the laboratory to separate proteins on columns used for their purification, but also on DNA molecules, in particular in the development of PCR tests. Unsurprisingly, demand has increased somewhat since the start of 2020.

The list is endless, and its content never ceases to amaze us. Recently, some even more miraculous properties of our heroic seaweed have been revealed.

At the beginning of 2021, a gene therapy based on a seaweed gene carried out by a Franco-American team at the Institut de la Vision in Paris enabled a blind man to regain his sight... The patient suffered from retinitis pigmentosa (RP), a degenerative disease that destroys the light-sensitive cells on the surface of the retina. Doctors used a protein that enables the seaweed to move by following the light. The gene for this protein was transplanted into the cells of the deep layers of the patient's retina, which then received other treatments. Today, he has not regained perfect vision but can see objects around him. A therapy based on a light-perception protein in a seaweed is therefore capable of partially restoring sight, even if this improvement has yet to be validated in the long term.

According to the director of the Institute that piloted the work, this approach goes far beyond the strictly visual framework and could also revolutionize the treatment of neurological diseases. Seaweed may not achieve the fame of Jesus making the blind see, but the potential for exploring its genomes and properties seems almost 'miraculous' all the same.

There is nothing surprising about the benefits of seaweed, since we share a common ancestor with the first complex life forms that emerged in the oceans over a billion years ago. The underwater world holds the keys to helping us hold on to our most precious commodity a little longer: our health.

All these valuable unexplored molecules constitute a reservoir of solutions for Western medicine, which needs to try something new to combat the obsolescence of historical antibiotics, to which pathogenic bacteria are becoming increasingly resistant. But also to deal with the emergence of potentially devastating new viruses and an ageing population.

Even if we've forgotten it for far too long, we are part of this marine environment. Unfortunately, this underwater world is not in good health itself. Most of its therapeutic treasures are disappearing amid general indifference, destroyed by water that is too hot, beneath tonnes of microplastics and other pollutants. It's high time to mobilize people and capital to save the marine biodiversity that created us. From our bacteria to our skin, the residual similarities between our bodies and seaweed are immense, and still unexplored. Besides the level of salinity, the plasma in our blood still has a composition very close to that of sea water.

We need to connect with the living world and understand that, in order to live healthier, we need to monitor not only our pulse but also that of the ocean.

Mastocarpus stellatus ('carrageenan moss' or 'false Irish moss') – a red seaweed from the North Atlantic which measures a few centimetres in length. It is often compared to *Chondrus crispus*, and shares many properties, particularly its high carrageenan content. More gelling than *Chondrus*, it has great potential in pharmacology against colds and flus. Its composition includes powerful antivirals. Its extracts are being studied as an alternative to plastic, but unfortunately it is not cultivable at present (its spores clump together in the form of crusts that cannot be removed from rocks). It contributed to the development of seaweed-based texturizers in the mid-twentieth century. It is still widely harvested today and used for gelling applications.

5

SEAWEED: NATURAL AND SUSTAINABLE INNOVATIONS FOR REPLACING PLASTIC, COLOURANTS, FOOD INGREDIENTS, FUELS OR TEXTILES

1914, Californian Seaweed Supports the War Effort

The Americans weren't the first to come up with the idea of making weapons out of seaweed. Napoleon had already thought of this in 1811 when preparing his armies to invade Russia. He had asked Bernard Courtois to study the possibility of using seaweed to obtain potassium nitrate, the basic element of gunpowder.

Historically, potash has been used in the production of soap, saltpetre and aluminium. It is obtained by heating wood ash mixed with water in large iron pots; hence the name 'potash'. Napoleon's chemists were convinced that, if wood contains it, it must be possible to find it in seaweed.

In fact, before them, in the eighteenth century, the Scots had already achieved this feat. Courtois did not succeed, but accidentally made a significant discovery. When handling these seaweed ashes, he unwittingly caused a strange purple smoke to appear, and became the first scientist to isolate iodine.

The use of this substance quickly became widespread after that, notably with the tincture of iodine already mentioned, which was widely used by Napoleonic troops for a purpose different from its initial demand.

Unfortunately for Courtois, although his discovery did not help save lives as he had been asked to do, it did save many and helped to heal millions of wounded on the battlefields and elsewhere during the centuries that followed. This 'failure' no doubt explains why Courtois died destitute and forgotten... A century later, at the dawn of the twentieth century, everyone sensed that the world was preparing to go to war again, even if no one really suspected that it would be truly global.

Almost all chemical products, and in particular potash, came from Germany, which exported them throughout the world. Without potash, there was no gunpowder, and therefore no weapons!

In the 1860s, Germany and its chemicals had made a major contribution to the American Civil War, which enabled the Americans to produce weapons. After 1910, the United States had to find alternative solutions. The huge forests of brown seaweed known as kelp were a great source of hope. At the time, the United States was the largest consumer of potash, buying half of German production. Part of it was intended to prepare for the war effort. Some of it was used by American farmers who spread it on their cotton, corn, potato and tobacco fields to improve their productivity. When war broke out, Germany announced an embargo on its potash production. The price of potash doubled in a few days. Investors funded research in California on a massive scale and were rapidly achieving very conclusive results with brown seaweed.

In 1915, the DuPont company, which at the time specialized in gunpowder, successfully invested in the processing of kelp and quickly began to industrialize production. Via its subsidiary Hercules Powder Company, in less than six months it built a huge factory on the seafront near the Mexican border, in Chula Vista. More than 150 containers stored a total of 3,200,000 litres of liquid kelp. The company operated around the clock and employed over 15,000 people who worked in appalling conditions.

Seaweed harvesting was carried out offshore with a minimum level of safety. The transformation process produced a pestilential smell; Chinese illegal workers carried out the most dangerous tasks; many lost their lives.

As the war intensified, DuPont developed the industrialization of its production methods and designed machines similar to maritime combine harvesters to collect this abundant giant kelp in the sea.

This seaweed grows back several times a year at a prodigious rate and provides all the necessary raw material. The machines were soon harvesting 500 tonnes every twenty-four hours.

The factory full of putrefying kelp was surrounded by barbed wire and guarded day and night by armed men who feared a German attack from Mexico.

Soon more than sixteen companies were producing potash and its derivatives from the large kelp plants. They all set up in the same geographical areas and collaborated with each other, creating a sort of gunpowder version of Silicon Valley.

The United States became one of the largest arms producers in the world, and at the same time engaged in a war that no one had imagined would last so long. While the productivity of seaweed remained relatively modest (less than 120 kg of potash per 10 tonnes of wet seaweed), the resource was abundant, and with the price of German potash having increased by 1,000% in 1916, the activity was very profitable. Especially as extraction processes were rapidly changing: fermenting liquified kelp to obtain not only potash but also a large number of other substances with multiple applications. Soon, kelp made it possible to produce iodine, acetone and other solvents or fertilizers quickly and simultaneously, for use in local industry or agriculture. One of the compounds of this seaweed contributed to the development of a new form of aspirin, distributed to soldiers at the end of the war to treat the symptoms of Spanish flu.

Once all the by-products have been extracted, what remains is a flammable liquid with strange properties that would much later come to be called 'biodiesel'. Quite ironically, as its value was unknown, the industrialists of the time generally burned it in order to dispose of it...

Another viscous substance is obtained in large quantities during these extractions: alginate.

Again, DuPont and their competitors saw this substance as worthless and missed an opportunity to sustain their seaweed business after the war was over. It was not until 1929 that the San Diego company Kelco understood the value and immense potential of alginate, which would soon invade our daily lives. Today, Kelco has built a multi-billion-dollar empire that continues to employ thousands of people and markets seaweed extracts and many other products around the world.

For much of the war, the Americans supplied the armies of the Triple Entente – especially England – with the chemical products they needed. California's kelp-processing plants also revolutionized explosives with innovative blends and detonation technologies that minimized inputs, maximized explosive power and facilitated transport and handling.

The victory was probably in part due to the quantities of kelp that were collected in the cold waters along the long Californian beaches.

After the war, the end of the potash embargoes and the low demand for weapons made it impossible to ensure the economic viability of these companies. One by one, they ceased their activities. The Chula Vista factory, the largest, closed in January 1919. The site of its location, now in the suburbs of San Diego, is still called Gunpowder Point. The employees of these kelp-processing plants were all awarded a 'War Workers Badge' by the government for playing 'a vital part in the prosecution of the war, second only to the part played by the man in actual contact with the enemy'.[81]

This World War I kelp industry remains to this day the largest seaweed industry ever created in the United States. Its development has led to the discovery of new substances that are widely used today in medicine, chemistry and food. The use of these substances, mostly as texturizers or gelling agents, has allowed the creation of vast production chains throughout the world and has shaped our consumer habits.

Today, seaweed is no longer used for warfare, but its potential is still huge. We hope that it will resolve certain imbalances on our planet, and thus contribute to maintaining peace by avoiding famines, climate refugees, economic disparities or the destruction of ecosystems.

Over a century ago, with much more rudimentary means than today, the 'war effort' allowed a change of scale that is unique in the history of seaweed in the United States. This acceleration of research and these large-scale investments have enabled the training of hundreds of qualified employees, the development of machines for harvesting, preserving and transforming a hitherto unknown resource, plus the development of elaborate transformation processes enabling seaweed to be broken down into valuable by-products.

These timely achievements are very much in line with the environmental and social emergency we are facing today. To accelerate investment and innovation in such a crucial sector, it is time for each of us to realize that global warming, world hunger and the loss of biodiversity are reasons that are just as valid as a world war.

Replacing plastic

In Émile Zola's novel *La Joie de vivre*, the author already had an intuition of this: 'The sea was a vast reservoir of chemical compounds; the seaweed was perpetually engaged in condensing in its tissue the salt which the water held in very diluted solution. The problem that they had to solve was how to extract economically from the seaweed its useful components.'[82]

Today, despite our lack of development in the field, the potential applications based on seaweed are so numerous that an attempt to list them would be like making an interminable inventory of unrelated items...

Some of them nevertheless deserve to be looked at in more detail either because they respond to major societal challenges or they've reached a current level of development that is worth mentioning.

Seaweed could soon replace plastic, which is a phenomenal challenge. Plastic is toxic and takes more than 700 years on average to degrade. Its use in food, textiles, construction, furniture and everything around us is increasing and has fuelled much of our economic growth. Over the past twenty years, the world has consumed more plastic than in the previous fifty years. In addition, this material emits over 60 million tonnes of CO_2 in its entire life cycle and therefore contributes significantly to global warming.

According to the United Nations Environment Programme,[83] less than 10% of the 8 billion tonnes of plastic produced since the 1950s has been recycled and we are expected to produce another 600 million tonnes per year by 2025. This production is mainly for single-use packaging. About 12% of this amount has been incinerated, at vast expense, while the rest was thrown into landfills or in nature. An increasing proportion of plastic ends up in developing countries, where no effective recycling system is in place.

Worldwide, 40% of plastic objects are thrown away after one month. Every year, an average of 10 million tonnes of these synthetic materials ends up at the bottom of the oceans. So today we have about 150 million tonnes of these microparticles floating in the sea, half of which come from food. Added to this is the equivalent of a refuse truck full of plastic that is dumped into the ocean every minute!

At the current rate, by 2050, the mass of these polymers in the ocean is expected to be 750 million tonnes, which would overtake fish. These 'plastic continents', made up of 'soups' of invisible microplastics and underwater accumulations of macro-waste, have been created at the confluence of ocean currents east of Australia in the Pacific, in the Indian Ocean and, to a lesser extent, in the North Atlantic.

These concentrations will obviously have dramatic short- and medium-term consequences on terrestrial and marine ecosystems and, no doubt, on our health. Not to mention that this material is mainly produced from crude oil, a non-renewable resource that will of course become scarcer and for which no alternative has yet been identified.

The food industry is the largest consumer of single-use plastic and is looking for the magic formula to replace this material, which is cheap to produce, lightweight, very strong, easy to transport and perfectly preserves the qualities of the food it protects from external bacterial attacks. Thanks to plastics, food can also be stored for much longer.

The search for a substitute has so far been fruitless. Big brands and international food organizations are therefore caught in a dilemma: their options are to drown the world in plastic or to completely deprive it. Plastic has become their main concern, and photos of turtles, birds or fish suffocated by plastic bags flood the media on a daily basis.

But halting the production of plastic without an efficient alternative would cause huge waste and losses in food production. This would lead to a sharp increase in food insecurity, as food would become scarcer and harder to transport to the people who need it. Not to mention large-scale food poisoning from meat or other sensitive products.

Here again, we must remain optimistic; there is no doubt that a range of solutions will create a new model for packaging and gradually replace the myriad uses of plastics in our society. In this respect, seaweed offers quite exceptional and – for the moment – largely underused properties. In fact, petroleum-based plastic comes largely from seaweed: dead seaweed that sank to the bottom of the ocean and sedimented over thousands of years to create oil.

New technologies are making it possible to considerably accelerate this process to obtain a healthier, recyclable or biodegradable and sometimes even edible 'bioplastic'!

Notpla,[84] a London-based start-up, has made a name for itself by designing a water bubble with a seaweed container intended to be ingested along with the contents. The plant and seaweed extract mixture is dried to obtain a powder which is transformed into a viscous fluid. When dried, it forms a plastic-like membrane that can be shaped in a variety of ways. The resulting water bubbles, ranging from 20 to 150 ml in size, are prepared at the user's premises in a machine about the size of a large fridge. The machine, connected to a water point, contains refills sold by the start-up. This produces dozens of water bubbles and allows people to drink a serving of water without using plastic cups. The packaging is tasteless and can be swallowed along with the water inside it, the principle being akin to the skin of a grape or cherry tomato. It simply disappears.

A few years ago, the organizers of the London Marathon distributed more than 30,000 bubbles filled with energy drinks, a welcome replacement for the usual plastic cups that are used to drink out of for seconds and take centuries to degrade. The model could be used for different products, from water to coffee, or alcohol for festivals, concerts, etc.

The Covid-19 crisis forced the company to focus on individual food packaging and to develop protective films and seaweed packaging to replace plastic. By the end of 2021, the first 150,000 seaweed takeaway boxes had been delivered to one of the UK's leading home delivery companies, and Notpla has now sold over a million boxes in six countries. Previously, the company worked with Kraft Heinz to replace mayonnaise and ketchup packets made from plastic. Notpla's packaging products respond to specific plastic problems identified across multiple industries including food, cosmetics, fashion and electronics, and could cost less in the long run.[85]

Seaweed is proven to be much better than many other grain-based alternatives that require fresh water and usually pesticides. Unlike PLA, a bioplastic synthesized from cornstarch, seaweed-based material biodegrades in nature within a few days without needing industrial composting conditions. The fact that it is edible truly demonstrates how easily nature can deal with it at end of life. And the resource seems abundant. According to the London-based start-up's calculations, 0.03% of the kelp available on earth could easily replace all the PET plastic used on earth. In late 2022, Notpla won the 'Build a Waste-Free World' category of the Earthshot Prize 2022: one of the most challenging competitions in the world, scouring the globe for the most inspiring and impactful solutions to repair the planet (thousands of competitors from across hundreds of countries).

Aside from kelp, there is also immense potential for producing valuable polymers using other types of seaweed, as demonstrated by the American company Loliware, who have launched drinking straws and cups made from red seaweed.[86] In Iceland, red seaweed has been used to make the first prototype water bottles. In France, a company from Marseilles[87] uses the green seaweed that swamp its beaches to produce a resin for manufacturing biodegradable cups, bin liners, packaging or meal trays. The number of start-ups around the world working on seaweed extracts as an alternative to plastic is growing steadily and is attracting the interest of major food brands, which are starting to discreetly launch pilot projects in their subsidiaries.

Outside of the food sector, seaweed-based plastic substitutes are sparking people's imagination. In France, a company recently won an innovation award for designing a seaweed-based biofilm that is intended to replace plastics for mulching fields for the horticultural and market-gardening industry. This biofilm is interesting for two reasons. It degrades naturally after a certain time due to light and rain, and, in doing so, it releases active ingredients that contribute to the growth of plants and the richness of the soil. The concept therefore combines mulching and the biostimulant effect, while eliminating waste, and has been supported by many scientific publications worldwide.[88]

In all these examples, the ability to process a sufficient volume of seaweed has yet to be demonstrated. In order to replace a product such as plastic, rather than 30 million tonnes of seaweed, we would need to produce several hundred million. Although wild resources are available, their large-scale removal is still logistically complex and ecologically dangerous. Moreover, the facilities for processing seaweed into biomaterials are still rare and very small-scale.

A change of scale and major investment will be needed to develop not just seaweed cultivation, but also processing. One of the challenges will be to avoid adding other chemicals. The reuse of marine plants in other co-products will also be a necessary prerequisite. This will provide an economic balance and avoid having to choose between producing a substitute for plastic or providing essential food to feed the planet.

The creation of relevant labels and standards will also have to be supported in order to truly warrant the investment effort made by these pioneers. The notions of 'recyclable' or 'compostable' currently displayed on products often remain vague because they are linked to various standards with varying levels of requirements. This vague attempt at standardization does not yet sufficiently define the highly effective environmental solutions that seaweed represents. Regulations for seaweed-based biodegradable packaging need to be created on the basis of sound principles and standardized across the world. When Notpla was founded, the directors were forced to declare their material as food, because there were no existing local regulations governing the use of seaweed as food packaging…

Textiles made from seaweed fibres

After some fine collective efforts to acknowledge the seriousness of plastic pollution, the recent health crisis has promoted the use of synthetic face masks, which now litter our pavements and will soon accumulate in the oceans. The carrageenan-based nasal sprays mentioned earlier appear to be excellent news in this respect.

Moreover, if we consider that a lot of the microparticles in the ocean also come from the washing of synthetic fabrics that slowly disintegrate in the washing machine and end up in our wastewater, we can see that using fibres made of seaweed is a worthwhile idea. Seaweed even seems much more appealing than cotton and other natural fibres. Cotton actually requires a large amount of water and contributes to the drying-out of the soil. It takes an average of 2,500 litres of water to produce a 250-gram T-shirt.[89] In 2016, 64% of the cotton grown in the world was genetically modified.[90] In addition, cotton cultivation is one of the most polluting in the world. According to the WHO, it takes up about 2.5% of the world's cultivated area, but consumes 25% of insecticides and 10% of herbicides.

These figures are worrying for the environment as well as for our skin microbiota. Seaweed-based fibres provide clear advantages here, because the soluble alginate can be woven in the same way as the fibres of terrestrial plants. Native Americans use seaweed extracts in textile-making to fix pigments. Mixed with other fibres, seaweed can be very easily dyed and can be used to create a wide variety of cloths.

Some forward-thinking designers are even considering using pure, unprocessed seaweed in their creations. These garments aren't destined for high-street fashion stores, but they are part of a creative approach to a new material that can play a significant role in changing our society's perception of marine plants.

But the use of seaweed should not be limited to haute couture. Like the cotton plant, a descendant of green algae, seaweed could provide quality fibres in large quantities if we work out how to process them. Seaweed-based textiles do not require any chemical treatment, allow good breathability for the skin and are odourless.

Certain clothes containing algae-based polymers also generate an active exchange of restorative substances between the fibre and human skin. The humidity of the skin allows a remineralization process that acts directly on our body in contact with the clothing, to repair cellular damage, reduce inflammation and soothe itching. The product is currently sold for runners who suffer from chafing, as they have the advantage of wearing close-fitting clothing while generating enough sweat to activate these properties. The T-shirt still costs around 200 dollars, but the concept is very innovative.

At the same time, a major retailer in Ireland[91] has recently started marketing sustainable underwear made from seaweed.

A growing number of shoe manufacturers are looking to incorporate seaweed into their production processes.

In Mexico, an entrepreneur has found a way to solve two of the region's major problems: plastic pollution and the proliferation of sargassum in the Caribbean Sea. He collects PET bottles from the oceans, mixes them with sargassum extracts and uses this to make shoes for the population.[92]

In Germany, a project for seaweed-based tampons is being developed.[93] Up to now, all tampons have been made from viscose, (sometimes organic) cotton and plastic. These materials contain chemicals such as dioxins and halogenated organic compounds such as glyphosate, which are concerning because they are used in the most absorbent part of the body and over a lifetime remain there for more than seven years. The environmental problems associated with their use are serious, particularly in terms of plastic pollution.

The idea is therefore to replace cotton and plastic with materials produced from seaweed. Besides the biodegradable aspect, the absorbent, antibacterial and anti-inflammatory properties of seaweed also make it an appealing solution.

Seaweed in art: from dyes, to cyanotype, to poetry

In the future, many other applications of seaweed will certainly make it possible to replace unsustainable or even polluting products. For example, there are some great recent initiatives concerning inks, both for standard and 3D printers.

In the Caribbean, sargassum, which has filled the sea of the same name, is compressed into high-density blocks – the natural glue of the seaweed making them compact – to create building bricks.

In Brittany, an inventor[94] has perfected a similar technique using beached kelp to obtain a very beautiful material that is twice as strong as oak.

In terms of dyes and paints, the stakes are also high. Paint is usually oil-based, and 95% of it is derived from the petro-chemical industry. This has significant ecological and health consequences. Also, seaweed can provide paints that are 98% plant-based, which are more locally sourced, odourless and much less polluting. Seaweed has long aroused great interest in this field for its 'antifouling' capability. In order to avoid being entirely covered with parasites, seaweed has developed a set of techniques in symbiosis with other organisms. Certain bacteria hosted by seaweed, for example, produce a substance that prevents unwanted guests from clinging to their surface. The same substrate is integrated into dyes and used to prevent the adhesion of foreign bodies to painted surfaces, particularly on ships or other marine infrastructure.

In construction and architecture, seaweed is also important. The famous 'seaweed houses' with their impressive roofs on Læsø Island off the coast of Denmark are actually made from *Zostera*, a type of seagrass. However, seaweed is also used. Here again, we must look to Japan for the finest examples. Since the sixteenth century, the technique known as *shikkui*, generally used to plaster large wall surfaces and the grouting of stones in Japanese castles, is composed mostly of slaked lime, natural plant fibres and the seaweed extracts *Gloiopeltis* (*fu nori* in Japanese) and *Chondrus ocellatus* (*tsunomata*) species. These are used as a water retention agent and improve the workability and elasticity of the plaster. From an environmental perspective, this method requires little maintenance, stores CO_2 and uses minimal materials. Finally, it is a 100% recyclable, non-toxic, fire-resistant product with good antimicrobial qualities. *Shikkui* is still used today in the restoration of Japanese historical monuments.[95]

Many sculptors are also exploring seaweed with a view to integrating it into their creations, and some specialize in it. A guitar-maker we met recently is looking to experiment with building a guitar out of *Laminaria hyperborea*, because he believes it has unique acoustic properties.

In 1842, a British botanist named Anna Atkins had the idea of photographing plants to create a herbarium. So she made impressions of seaweed using the cyanotype process. She placed dried seaweed on sheets of sun-sensitive paper, and as the papers reacted to the light, they became coloured and the silhouette of the seaweed appeared in white on a Prussian blue background. These seaweed prints were collected in a book, *Photographs of British Algae: Cyanotype Impressions*, which is considered the first book of photographs in history.

When it comes to poetry, there are writers the world over who have been inspired by seaweed. The Japanese poet Buson wrote this haiku: 'In the spring rain / Coming alive again / *Wakame* seaweed',[96] while Kito gave us this one: 'Green seaweed / in the rock hollows / No one remembers the tide.'[97]

The French poet Paul Éluard evokes a 'writing of solar algae',[98] and Jacques Prévert recalls: 'Like seaweed gently caressed by the wind / In the sands of sleep you stir dreaming.'[99]

The British poet Renée Vivien, who wrote in French, travelled extensively in Japan and always associated the woman she loved with seaweed imagery: 'We lay our bodies on your beds of dry seaweed'[100]; 'Never take again the bitter path of the shores / Where the seaweed has the slow swaying rhythms of a thurible'[101]; 'You are my perfumes of amber and honey, my palm, / My foliage, my cicadas singing in the air, / My snow dying to be haughty and calm, / And my seaweed and seascapes.'[102]

The famous nineteenth-century American poet Henry Wadsworth Longfellow wrote a poem called 'Seaweed', which explores the creative process and in which he describes: 'When descends on the Atlantic / The gigantic / Storm-wind of the equinox, / Landward in his wrath he scourges / The toiling surges, / Laden with seaweed from the rocks.'[103]

It would take too long to list seaweed's contributions to songs, so let's just mention Frank Sinatra singing: 'When you're in love, the seaweed smells just like roses…'[104]

While not questioning the amorous sentiment, this all brings us directly to the question everyone usually asks, when they hear about the potential uses of seaweed… Alas, despite its name, there are no known psychotropic characteristics attributed to seaweed. Its name could not be more of a disservice: it is not a weed, in any sense of the word.

The fact of the matter is, from cuisine to sculpture, to fashion, poetry and photography, we should explore every avenue to bring about this revolution and to connect our art, too, to this original creation that comes from the oceans.

Are algal biofuels on the horizon?

In addition to all these products, seaweed could also innovate in a sector which remains the biggest contributor to both global warming and planetary pollution: energy production. Although it only has a low calorific value, in the absence of other materials, seaweed has often been used as fuel through-out the world. Creating biofuel from seaweed by extracting lipids or bioethanol from sugars is technically possible and would offer a great alternative to fossil fuels.

The advantage of this energy source is that it has a lower carbon footprint. Algal fuel releases carbon dioxide when burned, in the same way as fossil fuels. But, as substitutes for plastic and other petroleum-based elements, an algae-based biofuel will not add carbon that has been trapped for thousands or even millions of years to the atmosphere.

There was significant enthusiasm for this biofuel in the early 2000s. Shell, Chevron, General Electric, Statoil and even Bill Gates invested large sums of money into the field at the time.

The US Department of Energy launched a multi-million-dollar programme involving US research centres and uni-versities. The resource seemed unlimited, untapped and, above all, did not compete with land-based food produc-tion. Unfortunately, the yield of seaweed to produce fuels soon proved to be very limited and the production costs were absolutely prohibitive. The techniques for converting seaweed into bioethanol are very different and much more complex than those used for terrestrial plants, because the structures of these types of seaweed are themselves more complex.

With our current state of technology, it does not seem con-ceivable that we can produce seaweed on a large enough scale to make fuel out of it, especially with regard to macroalgae, which are much less rich in lipids than microalgae.

These investments have nevertheless made it possible to develop maritime production techniques. Many companies in the United States, Norway and the Netherlands have been set up to produce biofuel from seaweed and have now become pioneers in the feed markets for both animal feed and biostimulants.

Moreover, while there is no indication that seaweed cannot become a competitive energy source in the longer term, its current development as a food resource may make this first potential even more complicated from an ethical point of view. As with cereal crops, the industry may have to choose between using seaweed to produce energy or using it to feed the population.

In the end, we should not lose sight of the fact that any large-scale production can become a problem for the balance of ecosystems.

However, we shouldn't be too definitive in our conclusions about this resource, which continues to surprise us.

Electric seaweed coming soon?

If seaweed could realize this potential, it would become the most rock 'n' roll organism on the planet! At a time when Europe's energy deficiencies are becoming clear, like guitars, seaweed could be electric. Electron transfer during photosynthetic activity has long been identified in microalgae. Microalgae are already used in the manufacture of certain solar panels, thus making it possible to reduce the use of toxic products. Unfortunately, these microorganisms produce only small quantities, and like the solar panels installed on our roofs, produce nothing when there is no light.

That's the point we were at, until a scientist went swimming in the Red Sea. Ever since Archimedes, it is proven that the most 'revolutionary' concepts have arisen during a prudent dip in the sea... This scientist from the University of Haifa tells us that while sunbathing on a beach, he noticed a rock covered in seaweed that looked like electrical cords.[105] He was already working on microalgae power and had the idea of using macroalgae to produce electricity. Eureka! The research led to a publication that was a shock to algae specialists, demonstrating the greatly increased electrical production potential of these marine plants compared to phytoplankton. Seaweed uses the energy of the sun to produce carbohydrates and biomass. This change therefore generates energy production. Logically, then, one might ask, why stop at seaweed? Why not study the electricity produced by other photosynthetic organisms such as terrestrial plants? But in physics, everything depends on the conductor. Seaweed is immersed in a highly conductive aqueous environment, unlike the air in which terrestrial plants live. While it is very simple to recover the electricity produced by these seaweeds in water, this is impossible to apply to other plants. At sea, these experiments were carried out on *Ulva*, or sea lettuce, which proved to be the most effective of the ten or so species of seaweed tested to date. Electrodes stuck to the blades of these *Ulva* powered a light bulb. Moreover, the electrical production of *Ulva* is 1,000 times higher than that of microalgae. Better still, even in the dark, this seaweed still releases 50% of its daytime production thanks to the use of carbohydrate stocks. To put it simply, let's say that seaweed has a sense of rhythm: it knows how to accumulate reserves during the day in order to continue producing wattage at night, whatever the weather. Even more efficient than photovoltaic panels and wind turbines, *Ulva* is a mini electrical power station!

And the story doesn't end there, because while these natural turbines produce power and sequester carbon throughout their lifetime, the resource is still valuable once they reach maturity. In the case of *Ulva*, the process of creating electricity seems to produce more electrons – and therefore more energy – than the seaweed needs to grow. So recovering its electricity does not damage the seaweed and only slightly alters its growth. We can therefore envisage using *Ulva* for its electricity, then cooking it for consumption or transforming it into packaging.

Phycocolloids: the seaweed extracts we consume every day

There is at least one recent industry which has been able to develop very rapidly and create immense production areas in countries without a history of growing seaweed.

An industry that has managed to build sophisticated global supply chains, while relocating the processing to consumer countries.

An industry which today is worth several billion euros and generates significant profits. In less than half a century, this industry has trained tens of thousands of farmers and introduced seaweed farming to the Philippines, Tanzania, Morocco, Malaysia, India, Madagascar, Spain and Tunisia. It has turned Indonesia – previously a newbie in this field – into the world's second-largest producer with almost a third of global production, far ahead of Korea and Japan, which have had centuries of experience.

The products of this industry have an unfamiliar name, even though we consume them several times a day: they are algae-derived hydrocolloids, or phycocolloids.

Hydrocolloids are texturizing gels for various products, mainly in the food industry. They are derived naturally and can be plant-based, such as pectin, guar or locust bean gum. They can also be bacterial, such as xanthan; or anionic, such as gelatine. But today, almost half of them are produced from the polysaccharides of marine vegetation, such as alginate from brown seaweed, and agar-agar or carrageenan from various red seaweeds. In seaweed, these complex sugars form the wall covering the membrane of each cell, which gives our sea vegetables the flexibility and elasticity to withstand the currents and waves.

In our food, these three very different substances with very specific uses are available in the form of multiple by-products with various qualities. They can be used as a texturizer, thickener, binder or gelling agent. These are essential elements of our food industry. These compounds – besides the fact that they are cheap – are odourless, tasteless and relatively unprocessed, cannot be absorbed by our body and are therefore also calorie-neutral.

Animal hydrocolloids were used as early as the seventeenth century and those of vegetable origin from the eighteenth century onwards. The industrial discovery of the almost miraculous properties of phycocolloids, and the ensuing commercial development over the course of the twentieth century, led to a large increase in production and a much wider range of applications. More recently, they have benefited further from the huge increase in demand for vegetarian or vegan products and the concern about animal products, following the outbreak of BSE (mad cow disease) of the 1990s.

These phycocolloids are therefore used today in the usual manufacturing process for ice creams, dessert creams, juices, sweets, preserves, meat, margarines, soups, terrines, sauces, pastries, pet foods and even certain beers. Some giants of the food industry admit to using it in 60 to 70% of their products.

It might come as a surprise that gelling agents are used in drinks, which by definition are liquid, but the use of these additives is in reality very variable and increasingly sophisticated.

In beer, they have a clarifying effect and limit the excess foam formed by proteins.

In canned meat, they are used to stabilize oil-in-water emulsions during preparation, cooking and storage, and to improve the appearance of the product.

Likewise, in pâtés, they are used as a substitute for fat and therefore have the effect of increasing volume and water retention, and making cutting easier.

In soft drinks, they significantly improve the texture and keep the flavour enhancers in suspension.

Some people will lament the processing of natural products. They are partly right… However, the growing use of these hydrocolloids is largely a consequence of our daily choices. Choices that sometimes differ from the principles we express. It is, after all, our appetite as consumers for aesthetically appealing, low-fat, low-sugar and even frozen food that has forced the industry to find ways of meeting these demands.

Freezing a product often necessitates protecting it from the cold by covering it with a layer of neutral gel. Removing sugars and fats destroys much of the original texture of food. And since most people prefer not to drink their meals, natural, low-calorie solutions had to be found to make our food look more attractive. These additives are also used for manufacturing toothpastes, creams, shampoos, shaving foams, soaps, deodorants, lubricants, polishes, etc. All these products would be almost as liquid as water without texturizers made from seaweed!

To be as comprehensive as possible, we should also mention their use in medical or pharmaceutical applications, the manufacture of fire extinguisher foams and even welding rods. In European supermarkets, these additives can be found on the labels of our products under codes ranging from E401 to E407.

Any child who has spent their holidays on the rocky beaches of the northern Atlantic has no doubt had the chance to pick up the red seaweed known as 'Irish Moss' (*Chondrus crispus or Mastocarpus*), dip it in boiling milk and see it turn to cream within a few minutes. The texturizer derived from this seaweed is called 'carrageenan' and takes its name from the county of Carrageen in Ireland, where it was used to make ointments and flans.

Although these structuring properties have been well known for centuries among coastal populations, major Western manufacturers only discovered this potential after the First World War. Until the 1960s, the natural sources of *Chondrus crispus* in Canada, France or Korea seemed to be sufficient to produce all the carrageenan needed. Agar was also wildly sourced in Chile. While there seemed to be no shortage of alginate supplies, given the size of the brown seaweed, it soon became clear that the same wasn't true for the much smaller red seaweed: the supply was quickly depleted by large-scale commercial exploitation.

In the 1960s, the food industry was concerned that its golden goose – with properties so highly valued by consumers – would disappear. Fortunately, red seaweed grows very well in tropical areas of developing countries. The manufacturers dominating the food additives market at the time invested in research into those red seaweeds, particularly the ones richest in carrageenan. Growing conditions were optimal in some areas and the main cost was labour, which turned out to be very cheap in those countries.

The extraction of hydrocolloids from red seaweed is actually relatively simple, since it only needs to be boiled in water then bathed in an alcoholic solution in order to precipitate the carrageenan fibres, which are then dried and reduced to a powder ready for use. In South East Asia and China, a semi-refined carrageenan is even prepared without using alcohol.

This production was therefore carried out with relatively low investments for high profit margins, achieved through selling these dried red seaweed powders to the factories of Nestlé, Danone, Unilever and others... No one would go so far as to say that it was the main objective of these private companies, but the development of this sector has largely benefited the resource-starved local economies.

The sector was established in what were often unusual circumstances, which we will come back to later. This development saw the emergence of pioneering researchers who realized the agronomic potential of these plants fifty years before anyone else.

Numerous mistakes were undoubtedly made because of the total lack of experience and the complex nature of the raw material. Many now accuse these companies of having acted too much in their own interests. If we put ourselves in the context of the time, we have to take into account that the levels of both social and environmental awareness have changed considerably since then.

Nevertheless, within a few years, these researchers succeeded in domesticating local seaweeds that were previously almost unknown. They set up a global industry using local production based on family structures that have been trained in innovative techniques.

Since 1990, the carrageenans sector has multiplied its production tonnage tenfold and the production of phycocolloids is still the only international seaweed-based industry that supplies all the countries of the world.

Although the types of seaweed introduced have often modified the local biodiversity and trophic balances in the cultivation areas, it must be acknowledged that a large-scale, non-polluting production has been created. This industry has also succeeded in bringing seaweed-based products into our homes and our daily lives.

Today, the raw material lacks genetic diversity and is suffering from the effects of global warming and extreme weather events. In some areas, seaweed won't grow, due to causes that are still not understood. Moreover, demand is maturing at a time when the Chinese market is flooding the world with its cheap texturizers without any regard for Western ethical requirements.

This experience is an interesting case study that can teach us valuable lessons about the path to establishing a sustainable sector.

The first lesson to be learned concerns the supply of the raw material and its production volume. If the types of application such as plastics, textiles and especially biofuel, mentioned above, are developed, they will require a very large supply of raw materials to meet the demands of already well-established industries and the needs of a rapidly growing population.

Here again, the problem arises how to cost-effectively domesticate, and then cultivate, large volumes of different types of seaweed. Finding the space to produce them, store them, dry them and the best way to transport them.

It is quite incredible to think that a product such as alginate, which has been used for almost a century and in so many products consumed around the world, is based entirely on a wild supply source harvested at low tide. Here is a billion-dollar artisanal market with an annual growth of 5%,[106] yet no one knows how to harness the production of its raw material.

Seaweed biorefineries

The other huge weakness of the seaweed sector, but also its greatest potential, is the fact that this raw material is only partially used, due to the current state of our scientific knowledge. In order to build a large-scale seaweed industry that goes beyond direct supply, it will be essential to successfully break down our plants into several by-products that can be used in different markets.

The immense carrageenan industry in the Indian Ocean uses its seaweed solely for manufacturing texturizers. This model was profitable a few years ago, but now the selling price and yields are falling, while significant investment in research is needed to improve germ plasm and seedings.[107]

Unfortunately, this industry continues to dispose of the residues of its raw material after extraction. It therefore rejects – in the form of waste – valuable molecules that it could sell at a decent price. This recovery of by-products is possible, but involves studying the processes needed to extract and market them. And this chapter has only just begun.

France has some fine forerunners in this field. The Algaia company in Normandy is one of the pioneers in extracting and marketing the co-products obtained after the extraction of texturizers. The brown seaweed alginate that it sells represents only 25% of the dry seaweed it derives from. Previously, the soluble part of what remained was mostly dumped in sewage treatment plants and the solid part was spread on the fields. However, this remaining 75% is a significant source of valuable compounds.

Algaia is now demonstrating that the remaining alginate residues can be recycled into bioplastics for much more attractive margins than pure alginate.

In addition, the active ingredients in this 'waste' are an effective biostimulant for arboriculture, viticulture or market gardening, while others can be recycled into cosmetic products, in particular into prebiotics for the skin microbiota, to prevent dry skin and so on.

In addition, carbon residues have the potential to revitalize soils and enhance bacterial activity within them. And all this is what we call 'waste' in seaweed! It would be like rejecting the wholegrain part of cereals and just extracting the starch.

A seaweed biorefinery is the key to success, but there are numerous difficulties when it comes to designing it cost-effectively. The financial balance is very complex and significant upstream investments are needed. Furthermore, creativity is required, because the current extraction processes for alginate or carrageen cause too much damage to the residual active ingredients – such as the proteins or nutrients – for them to be exploitable at reasonable costs in the future. At the same time, extraction residues containing polluting chemicals still pose problems when it comes to commercializing these seaweed co-products.

We will therefore have to continue to look for other, more precise, more efficient and 'greener' means of extraction, because the economic viability of seaweed depends largely on our ability to better extract, understand and enhance all the by-products that this raw material can provide.

This type of operation can help create a more profitable, ethical, redistributive and investment-capable sector in the years to come. It will also be a sector that is more resilient and able to easily reorient its various production streams according to the hazards of an international market which changes rapidly due to frequent innovations.

Domesticating seaweed, increasing production and extracting valuable compounds will involve constant work and collaboration between the worlds of research and industry, which, here as elsewhere, have too often taken divergent paths. But what is at stake is worth it.

So that in every object, in our clothes, our dyes, our art, our furniture, our energy, our food, we stop building our society on a polluting resource that has been dead for millions of years. Through a closer connection with the oceans, we could reap the benefits of a sustainable, living, versatile, clean, biodegradable, local, carbon-negative resource, about which there is still so much to discover.

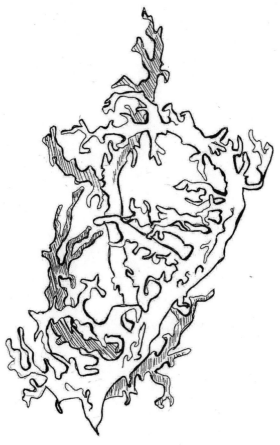

Kappaphycus alvarezii (elkhorn sea moss, *guso* or *Euchema cottonii*) – a species of red seaweed, which can also be green or yellow in colour. It is one of the five most cultivated seaweeds in the world. It is named after Vic Alvarez, the Filipino researcher who helped domesticate it and discover its commercial properties in the 1970s. It grows very fast and doubles its biomass in a couple of weeks. It lives mainly in the Indo-Pacific region and in South East Asia. It is one of the most significant sources of carrageenan and helped to create the world's largest red seaweed industry at the end of last century.

6

SEAWEED CULTIVATION: AN ETHICAL SECTOR TO COMBAT POVERTY IN COASTAL POPULATIONS

Women and Seaweed
Take Power in Zanzibar

Zanzibar is a group of coral islands edged with long white sand beaches, lapped by shimmering turquoise waves. It was here, at the tail end of the 1980s, that a unique socioeconomic experiment began, using seaweed to support the then flagging local economy. Even better, this new source of income mainly benefited women and gave them some autonomy in a country with very patriarchal traditions. This unique story is usually linked to an equally unique researcher who devoted many years and energy to seaweed in Zanzibar and is still very active in the field: her name is Flower Msuya.

Flower grew up a six-hour drive from the sea, in a farming family, in a small village on the slopes of Kilimanjaro. Nothing could have predicted that she would become the pioneer of seaweed cultivation on a remote island... She first studied botany at the University of Dar es Salaam, where she became passionate about seaweed thanks to her teacher, Keto, a visionary who wanted to set up a red seaweed industry in order to extract texturizers and export them to the West. Apart from being used in deworming treatments, seaweed had never been part of the local culture of Zanzibar.

Flower recalls how, when she told her family that she wanted to study seaweed farming, they were surprised to find out that there were also 'trees under the sea'. The young woman went off to study in Finland and then in Israel thanks to the aid of local governments and the WWF. Back in Zanzibar, she applied what she had learned and developed algaculture with the help of Keto. She divided her time between the Institute of Marine Sciences and the beaches of Zanzibar, where she supported women's work. Together, they worked on cultivating seaweed from cuttings attached to ropes and placed in the sea and shared the best techniques for growing and preserving seaweed. They also tried to identify the best strains and favourable locations to grow them.

In the early 2000s, Flower created and led the Zanzibar Seaweed Cluster Initiative, in which she brought together producers, industry, government and academia within the same structure.

Today, seaweed farming employs over 25,000 people in Zanzibar, 80% of whom are women.[108]

This new culture is also developing on the mainland in Tanzania. Flower has even travelled to neighbouring countries to spread the word. Red seaweed has therefore become the third-largest source of income for Zanzibar after tourism and its traditional export, cloves.

Seaweed represents 25% of the island's GDP and 90% of its maritime product exports.[109] Tanzania has become one of the major seaweed-exporting nations and the only country outside Asia to be in the top ten producers of cultivated seaweed. In volume, Zanzibar reaches a quarter of the production of Japan![110]

Zanzibar's history has always been unique in every way. The place has experienced Persian, Portuguese, Omani and English occupation, interspersed with short periods of independence. The rulers of the day always saw it as an important trading post, ideally positioned at the crossroads of the Arab, Indian and African worlds.

In addition, the clove tree grows magnificently there and contributes to the wealth of those sun-drenched islands. The labour required for its difficult cultivation also made the archipelago a major hub for slavery and later for the trafficking of gold, spices and ivory until the end of the nineteenth century.

In 1964, at a time of African independence and after years of conflict with the English and then between the two neighbours, mainland Tanganyika finally merged with the island of Zanzibar. Each of the two regions gave its first syllable, to create the United Republic of Tanzania. Unfortunately, this country has not been able to capitalize on its history or its position, and today has one of the lowest human development indexes in the world (198th out of 228...).[111] Recently, the development of tourism has led to the modest beginnings of economic growth. But as soon as you step outside the luxury hotels, life is hard in Zanzibar. And even more so for its women...

The massive production of cloves in Madagascar drove prices down. The coral lands do not allow the cultivation of cereals, while on the coasts, the fisheries are increasingly meagre. Flower recounts the daily lives of women who were limited to doing only household chores, were not allowed to have a job, could not go out without their husband's permission and had no access to money, education or free speech.

That was until 1989, when seaweed came along...

To understand this story, we need to go back to the early 1960s and to the University of Hawaii, where the research centre of Maxwell Doty, one of the most renowned phycologists of his time, is located. Maxwell worked with the Chinese researcher C.K. Tseng – who we will come back to later – and then specialized in red seaweed from South East Asia. At the time, the scientist was working in partnership with a large US company, Marine Colloids Inc., to try and find a solution to the limited wild resources supplying carrageenans.

The growing demand for Canadian *Chondrus crispus* was prompting significant harvesting. This wild resource was now threatened with extinction. Prices skyrocketed. The food industry travelled the world in search of what it called the 'holy grail' of carrageenan cultivation. If *Chondrus* could not be cultivated, the producers at the time hoped to find another cultivable red seaweed capable of producing the prodigious substance. Maxwell Doty believed that this seaweed was hidden in the western Indian Ocean. And he would be proved right in the future.

In 1966, the researcher, then in a relationship with a woman of Filipino origin, discovered a strange red seaweed at a market in the south of the country which he decided to study. The seaweed did not yet have a name but could grow to twenty times its size in two or three months and produce very high-quality carrageenan. Maxwell shared his research with some of his best students. One of them, Vic Alvarez, was also Filipino. A keen researcher, he would later head up the country's enormous red seaweed industry. Maxwell decided to name his new

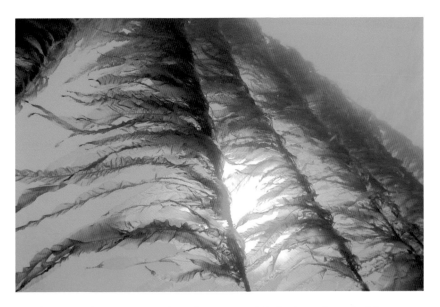

Atlantic wakame seaweed at the Seaweed Solutions farm in Norway.

Seaweed farming in Xiapu, China
(photo: Shutterstock).

Himanthalia elongata, also known as
sea spaghetti, in Brittany, France.

The illustrator with a 'Mwani Mama' working on a red seaweed
(*Eucheuma*) farm in Paje, Zanzibar.

Author visiting a seaweed farm in Lofoten Islands, Norway.

40-metre-high *Macrocystis pyrifera* forest in Chile.

Ascophyllum in Oban, Scotland.

Author with photobioreactors for scientific
research in Roscoff Marine Lab, France.

Workers drying organic seaweed (cabbage)
in Khanh Hoa province, Vietnam (photo: Shutterstock).

Woman in Bizerte, Tunisia, seeding
red seaweed in tubular net.

Author and *Durvillaea* in Tasmania, Australia.

Author in Lesconil, France, visiting Algolesko,
one of the largest seaweed farms in Europe.

Giant kelp (*Microcystis*) harvested in
Kelp Blue Offshore farm, Namibia.

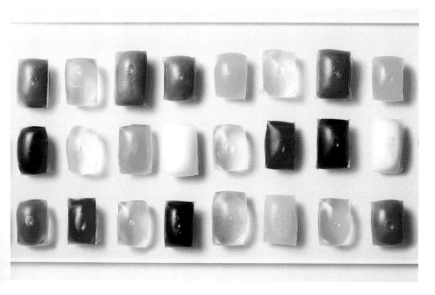

Notpla Oohos, edible bubbles made from seaweed.

A range of seaweed-textiles by the designer Violaine Buet
(photo: Pierre-Yves Dinasquet).

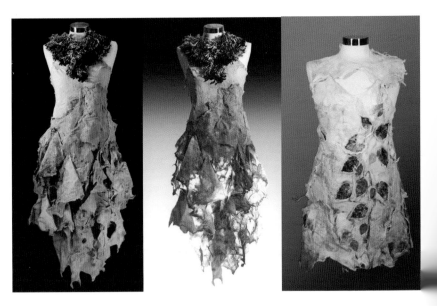

Seaweed dresses by Runa Ray.

red seaweed *Kappaphycus alvarezii* in his honour! Another of his students, Keto Mshigeni, would become Flower Msuya's famous mentor. The world of seaweed science is a small one!

After eight years of relentless research, the researcher from Hawaii and his colleagues finally managed to domesticate *Kappaphycus*. In the quest for carrageenans, Maxwell and his teams benefited from large budgets allocated by manufacturers in the US (Marine Colloids), France (ex-Sanofi) and Denmark (CP Kelco). These investments have proven to be worthwhile and have enabled the three companies to generate 90% of the world's 'red gold' production. These companies became part of various international groups that still dominate the sector today.[112]

A few years later, Maxwell domesticated an alga of the same family, *Eucheuma denticulatum*, which provided a more flexible carrageenan for softer gels for other, less remunerative applications. Within a few years, the two sisters, *Eucheuma* and *Kappaphycus*, became the pillars of a fast-growing global carrageenan production. An industry was born!

Despite civil wars and typhoons, the development of the cultivation of these two red seaweeds was impressive in the Philippines, especially in the south, in Tawi Tawi, an island in the Sulu Archipelago. The sector's large Western industrial companies, who were buying up the country's seaweed production, soon gave up on the idea of an integrated sector and relied instead on more flexible family farming systems. The villages have populations that are available for work, knows the sea and whose hourly labour cost is low. At the same time, the investments needed for this new culture are marginal in terms of both training and equipment.

There are attempts to develop this model in Asia, but the precious seaweed is very temperamental. Its favourite places to hang out do not seem to follow any logic. So Maxwell and his followers would travel the world, trying to find the most suitable beaches for their new treasure. They considered the seas of more than forty countries: Brazil, Djibouti, Madagascar, India and especially Indonesia, where, finally, the conditions seemed to be ideal. It was a total success, and the cultivation of seaweed was established on a long-term basis. The market was flooded, thanks to the arrival of this prodigious resource believed to be infinite. The increase in supply accelerated research and led to the sophistication and diversification of applications derived from these red seaweed polysaccharides. Nearly fifty years later, carrageenan is still the most important texturizing agent used in the food industry and still comes almost exclusively from the two species discovered by Maxwell and his team.

Alas, Maxwell Doty came to a less glorious end. He returned home to Hawaii and took his favourite red seaweeds with him to continue studying them. But in 1996, at the very end of his life, some of his seaweed escaped and colonized the surrounding waters. These escaped *Kappaphycus* were then accused of contributing to a sudden deterioration of Hawaii's corals, on which the island's ecosystem depends. The seriousness of this degradation and the actual harmfulness of the professor's seaweed are still subject to debate.

The fact remains that in the United States the name of Maxwell Doty is still linked to an environmental scandal which has largely erased this scientist's visionary contribution. 'Trust is gained in drops but lost in litres,' said Jean-Paul Sartre.[113] This is even more true in the ocean.

By the 1930s, a relatively small trade in wild red seaweed had begun in Zanzibar, to Europe and America. Unfortunately, the resources ran out and the modest sector disappeared in the late 1960s. So Keto Mshigeni used the experience he'd gained with his teacher to recreate this red seaweed market in 1989, but this time by cultivating it. Flower soon joined him in the adventure. The early days were laborious because the endemic Tanzanian seaweed species did not produce enough carrageenans to be profitable. As the regulations of the time were less stringent than they are today, red seaweed from the Philippines were essentially imported legally.

This magic seaweed reproduces asexually, and so can be cut into pieces and have cuttings taken. Each cutting is capable of doubling in size in two weeks… Suffice to say that a suitcase of seaweed discreetly brought back from Manila is enough to create a new sector in Zanzibar within a few months. Supported by large companies from the West, the success was rapid, but the genetic heritage of seaweed is weak and the resource is not local. Although it is always easy to judge in hindsight, these two factors carried the seeds of the future problems that this cultivation would encounter. The private companies of the time had the vision, took the financial risks and implemented the cultivation. It is hard to blame manufacturers for not being able to better anticipate problems that took almost twenty years to arise.

At first, the results were convincing. Soon there were more and more trials in Zanzibar. The investment was minimal for the local populations: planting a few wooden stakes in shallow water, sowing cables stretched between these poles and letting nature take its course, following the advice of Keto and Flower. After a few weeks, all they had to do was collect the material and dry it carefully on the beach, avoiding the accumulation of sand. But the process requires patience and precision. And the volumes must be significant in order to make a profit.

At first, the men on the island tried it, but soon found it too time-consuming, delicate and unprofitable. They preferred to go back to fishing or tourism to have a few pennies in their pockets at the end of the day. But catches were becoming scarcer. Even if they worked hard, it was becoming difficult to earn a living.

So the women took over the reins. Flower tells us how, in the beginning, conflicts broke out among communities with very ingrained traditions. Husbands threatened to divorce their wives if they left the house to work on the beach. Therefore the first ones sowing the cables at sea were single mothers and divorced women.[114]

Very quickly, economic reality won them over and principles yielded to the income these pioneers were making. *Mwani* is the word for seaweed in Swahili, the regional language. 'Mwani is money' quickly became a popular motto. The women of Zanzibar began to earn a living and change their status. These 'Mwani Mamas', as they are called, acquired a certain independence and made their voices heard. In the early years of this cultivation, the island saw an explosion in sales of radios, bicycles, kitchen tools and especially kangas, the traditional fabric worn by women in Zanzibar.[115] Some of these women have become veritable entrepreneurs and have managed to pay for the education of the next generation. Others have travelled to share their knowledge, in particular to Kenya and Mozambique.

Seaweed farms are now visible at low tide, all along the coast of Zanzibar's islands. The Mwani Mamas are hard at work most of the day, against this vivid azure backdrop, attaching the lines and removing urchins and other unwanted organisms from the surrounding area. Walking down the beach, seaweed can be found everywhere, fluttering in the breeze, attached to 'dryers' fashioned from coconut tree branches.

Flower, often with her feet in the water, accompanied this movement and, in between preaching, tirelessly continued her research at the Institute of Marine Sciences. However, this research often turned out to be frustrating. Because not only did no one manage to improve the local strains in order to make them more effective, but of the two seaweeds imported from the Philippines, only *Eucheuma*, the least profitable, had satisfactory yields. One after another, attempts to grow *Kappaphycus* have failed.

To this day, this seaweed remains a mystery and no one can really explain why it grows in one place rather than another. To overcome this competitive deficit, the focus was on the quality of the product, which made it possible to serve the more demanding Western markets. Unfortunately, the sector was soon overtaken by the combined and well-known effects of seaweed's genetic poverty and climate change. Over a period of thirty years, surface water warmed by an average of one degree, which reduced yields and encouraged the development of diseases or 'parasitic' organisms. As a result, Flower took the lead in creating the first organization to bring together the various stakeholders in the seaweed sector in Zanzibar. In the early 2000s, they worked together to develop ways of making better use of this seaweed by transforming it on site.[116]

The growth of tourism on the island created a market in hotels for local products. So the organization then developed soaps, creams, juices and other products based on red seaweed and coconut.

Rudimentary processing tools were acquired so that women could process and sell these products without intermediaries. Although the volume remains marginal, around 1% of production, the margin is significant. As is often the case in Africa, with very few means, a big difference can be made. In the village of Paje, where the seaweed cultivation first started, women have banded together to build two schools with the profits from the sale of these products. In the same village, a few young entrepreneurs left their careers as investment bankers in London to come back to Africa and develop a patent for extracting precious skin care substances from these red seaweeds. They also created a 'Mwani Center'[117] which offers an ecotourism visit to their seaweed farm and introduces the various stages of processing.

On Pemba Island, which now represents the majority of the national seaweed production, a very ambitious processing plant is about to open in collaboration with two Indonesians firms. Flower still acts as a scientific consultant on the project.

These are initial steps towards systemic change which make it possible to re-localize the added value of the product while at the same time freeing it from the purchasing monopoly of manufacturers and the volatility of market prices. This new model is unfortunately still very marginal on the island today, and the sector faces many challenges. Carrageenan seems to have reached maturity and is experiencing variations, with disastrous consequences for small producers.

Moreover, the emergence of massive new competitive production zones requires a very high level of efficiency that is rarely achievable with village organizations.

Finally, serious accusations have recently been made against certain types of over-processed and degraded carrageenans that may be contributing to the development of cancers. This latest news led the European authorities to impose a maximum level of 5% for carrageenans in food products.

Today, the seaweed industry in Zanzibar is suffering, but Flower and her peers hope to gain the support of government and international organizations to change the cultivation model. Changes are already underway: coastal populations are learning to work in waters further from the shore, with tube-shaped nets that enclose the seaweed. These new, deeper and therefore colder zones are a measure to counteract the warming of surface waters. They make it possible to double yields and sometimes to grow *Kappaphycus*, which is more profitable than *Eucheuma*. But they also involve the use of boats and the obligation to know how to swim. Most women in Zanzibar are unable to swim or handle a boat. Flower tirelessly organizes new courses, which now include swimming and boating lessons.

The sector and the local research organizations are aware that they also need to work on the genetic quality of this seaweed and identify strains that are more resistant to warming waters without giving in to the temptation of genetically modified organisms. Although effective in many respects, such an option at sea would be very dangerous as it would be totally uncontrollable. There is no doubt that this subject will give rise to some interesting debates in the world of seaweed in the future...

The Tanzanian sector has also set itself the objective of developing domestic processing capabilities and adding value to its product locally. It could also create and promote end products that are less sophisticated, less processed and which perform more modestly but could avoid having the E407 label

(European code for carrageenan-based additives), which has acquired a negative and chemical connotation. By putting an end to extractions with sodium hydroxide or precipitations with alcohol, and getting back to the basic powders of *Chondrus* – which our great-grandparents used to make flans[118] – the sector could thus have an ethical and 'natural' label reflecting the value of its seaweed, which would better meet the demand of the Western market.

Tanzanian phycologists are also considering the option of turning to other seaweed. Producing agar from other types of endemic red seaweed appears to be possible, although still complex.

Some also suggest that local people should be educated to eat and cook seaweed so that it can be appreciated as food. Seaweed grown in Zanzibar has been part of culinary traditions for centuries, but very little effort has been made in this direction since 1989. This is also changing gradually with recent initiatives led by young women from mainland Tanzania and aimed at promoting seaweed as a healthy and nutritious food for the local population.[119]

The development of animal aquaculture raised in symbiosis with seaweed could also prove interesting, because it is regenerative. The Food and Agriculture Organization (FAO) and South Korean government funds have recently financed the development of a hatchery and training programme for the cultivation of milkfish (*Chanos chanos*), sea cucumbers and small crabs that will benefit from the presence of seaweed, and vice versa.[120]

Flower collaborated in the development of the Seaweed Manifesto in 2020 and was invited to speak at the release of the document on the fringes of the UN General Assembly the same year. She was also invited to share her experience at the launch of the Safe Seaweed Coalition organized in partnership with the *Financial Times* in 2021.

When she's not with the women growing seaweed on the beach, or in her laboratory working on new strains of seaweed, Flower sits on the steering committee of this global forum alongside her Japanese, Korean, European and American counterparts. All of them are working to represent this young seaweed sector in its efforts to standardize but also to promote research by attracting investors.

Today, 200,000 people in Tanzania still live indirectly from the income generated by seaweed. It is therefore crucial to continue supporting this pioneering industry in Africa. It has the qualities to attract impact investment from around the world, which will enable its renewal and growth.

The seaweed experience in Zanzibar offers us a perspective for Africa and shows us the pitfalls to avoid. Although seaweed farming is a real challenge, it holds the keys to tackling social justice and contributing to a fairer model in the future.

Seaweed geopolitics: a new model for coastal populations

Despite its size and its essential role, the ocean is largely overlooked by the economic and financial world. The United Nations has defined 17 Sustainable Development Goals (SDGs) intended to articulate our society's strategy and investments in order to overcome the ongoing crises. The

14th goal covers all subjects related to 'life below water'. This SDG is by far the most underfunded since, according to the OECD, the ocean only managed to attract 0.01 % of development funding and 0.5 % of philanthropic funding linked to these objectives.[121]

However, today, having reached or even exceeded the limits of our natural resources, along with strong demographic growth, we are being forced to consider new solutions. Populations ravaged by poverty are amassing on the coasts, while coastal resources, primarily fishing, are inexorably declining. Marine biotechnology is an innovative response to this socio-economic tragedy. The seaweed revolution will be complex, but it offers ecological solutions and new market opportunities for communities in great need.

This revolution already began in Asia during the second half of the last century. In 1950, the seaweed sector was almost non-existent. It is now worth 13 billion euros worldwide and has an annual growth rate of 8%, among the highest in the food industry. Seaweed is produced in 48 different countries on plots ranging in size from 1 to 1,000 hectares and is exported around the world. The sector employs millions of people directly or indirectly. In Asia, this transformation has been split between the north and the south according to two distinct geographical zones which have very different operations and objectives.

The northern zone includes Japan, North and South Korea, and China. We have seen that seaweed occupies a major role in the culinary traditions of these countries; it represents a profitable, fast-growing industry, with the main objective of feeding the population. Cultivated seaweeds – mainly brown and some red – are not very processed and are mostly used in their natural state. They are mainly sold in domestic markets.

There is very active innovation, and export of these seaweeds correlates with these countries' culinary traditions flourishing within the globalised catering industry.

The southern zone includes the so-called 'Coral Triangle' countries (the region is exceptionally rich in coral reefs) and includes the Philippines, Indonesia, Malaysia and a few surrounding islands (Solomon Islands, Papua, etc.). These countries do not have such strong traditions related to seaweed. Only tropical red seaweeds are grown there, but they account for a third of global seaweed production. The sole purpose of these raw materials is to be exported and processed, in particular as texturizers sold to the European, American and Chinese markets.

The only thing these two blocs have in common is the difficulty of dealing with the consequences of climate change and the warming of surface waters.

The model of edible seaweed in North Asia

In all the countries of the northern zone, seaweed is an institution, largely contributing to the nutritional autonomy of the countries concerned. It is the equivalent of major cereal crops in the West and benefits from organized representation, a precise governmental strategy and highly active scientific research and technical assistance. Its development has led to the creation of large private sector leaders that have diversified their products and now have a vision that guides the development of the industry. The political stakes are high for these sectors, and they employ millions of people.

For a long time, these food resources were only found in the wild and their availability was considered to be infinite. Then in the late twentieth century, wars and rapid population growth in these regions suddenly revealed the limits of the ocean.

So cultivation was therefore put in place. The main challenge for these countries today is finding new available space. The combined effects of warming and water pollution drastically reduce the areas suitable for farming seaweed.

Discreetly but actively, North Asian countries are acquiring stakes in other countries, particularly in the West, to grow a crop that they control, so as to supply their domestic markets.

Japan, the most emblematic country of the first bloc, is also one of the first to have truly developed the cultivation of seaweed. Seaweed is highly respected in Japan. From Hokkaido to the islands of Okinawa, seaweed fields are ubiquitous, and the different climates offer a wide variety of species. It is used in cooking in extremely sophisticated ways. There, algaculture remains traditional and small-scale. It is essentially manual and has proven to be very efficient. However, the seaweed sector suffers from a lack of interest from the younger generations and is struggling to find new operators. The Land of the Rising Sun exports very little seaweed but imports a third for its consumption. The latter figure has been rising sharply, particularly since the Fukushima nuclear disaster. If it is to carry on supplying its population, Japan will have to review its mode of operation in order to modernize its image and develop technologies that allow it to be more efficient.

South Korea has already begun this transformation. The third-largest producer of seaweed in the world, it currently produces four times more than its Japanese neighbour and has quadrupled its production in twenty years. It has ideal coastlines for cultivation, with *wakame*, *kombu* and *nori* accounting for 97% of production. The Land of the Morning Calm is now the largest exporter of *nori* in the world, far ahead of all others. Its production supplies the domestic market and floods the rest of the world. The sector is already largely mechanized and modern. It takes advantage of large-scale automated systems and boats specially created for seaweed cultivation. In South Korea, this sector is critical in terms of employment and is a real source of national pride.

Besides over-the-counter drugs, personal hygiene products and cosmetics, Korea – like Japan – does not really seem to consider seaweed for applications other than food, as this market already provides the industry with the necessary outlets.

It might seem surprising to find North Korea among the top five seaweed powers. Yet this industry is essential and allows this country to feed its population despite the embargoes in force. Its production exceeds that of Japan and is twelve times that of France!

Satellite images clearly show large seaweed farms at sea and increasing numbers of tanks on land. Seaweed also represents an opportunity as an energy source, in the face of international sanctions. Many indexes show that this resource is being used to produce biofuel. Marine plants are a major strategic asset that allows North Korea to maintain its policy of isolation.

Its massive neighbour China is the largest producer of seaweed with 60% of world production of all species of seaweed combined and plays a pivotal role between North Asia and the Coral Triangle. China is more similar to the former in terms of its traditions and the importance of seaweed in its diet. Sanggou Bay, across the water from Korea, is covered by 130 square kilometres of seaweed farms, sometimes integrated with oyster farms.[122] These dozens of farms produce almost a million tonnes of seaweed in this particular area. All along the coasts, huge areas of sea farming are helping to reduce agricultural and aquaculture pollution in the country's bays. Countless processing plants are dotted along the coastline and employ millions of workers. The techniques are sophisticated, born partly out of close collaboration on the subject with researchers in the Soviet Union during the Cold War. The Kremlin recognized sea farming's great potential for ensuring the bloc's food and nutritional independence.

According to China's Seaweed Association, three million people work in the seaweed industry or in seaweed production. This figure rises to five million if indirect jobs (transport, laboratories, etc.) are included.

Exports of seaweed from the Coral Triangle

In this region of South Asia, the sector was artificially created in the 1960s by Western manufacturers looking for new sources of raw materials, while these developing countries were looking for new sources of income. Production is currently facing numerous difficulties in a slow-growing international market with a constantly increasing number of new entrants. The region has a large amount of natural potential that is still mostly underused for the production of tropical red seaweed, since it has 70% of the available coasts on the planet within the equatorial zone (located at a latitude of ±10 degrees from the equator).

The production structures are fairly atypical; often family-run, they do not have a clear hierarchy and remain very traditional. The activity promotes social ties and is inclusive with, in general, with tasks distributed fairly equally between men and women. Seaweed cultivation is often a part-time activity, in addition to fishing or other aquaculture, but it takes place all year round, with eight harvesting cycles per year. It is usually done by hanging seaweed shoots on ropes stretched over bamboo rafts that cover many kilometres of coastline. The women take care of the tasks carried out on land, while the men work more often in the water. Permits and access to cultivation areas are often facilitated by the authorities in accordance with local traditions and therefore aren't hindered by the Kafkaesque complexities encountered in the West.

Initially, seaweed farming was a very lucrative business. During the 1970s, the revenue generated per hectare was five or six times higher than the minimum revenue on land for the same area.[123]

Moreover, the countries in the Coral Triangle have succeeded over the years in developing processing activities that have enabled them to re-localize part of the profits.

Major Filipino and Indonesian companies have emerged in the production and processing of carrageenan and are now competing with the West in regional markets. In the two main countries (the Philippines and Indonesia), there are high political stakes for the sector. Some people involved locally consider the official production figures to be largely erroneous. According to them, the figures reported by the countries correspond mainly to national commitments but in reality would be almost three times lower.

Cultivation in this triangle began in the southern Philippines, in the Tawi Tawi region, in 1974, thanks to the discoveries of Max Doty. All the species of red seaweed found in the Coral Triangle and elsewhere originally came from this country, which is still the fourth-largest producer in the world. Seaweed is an essential resource there and employs around 150,000 people.[124] It is also traditionally part of the nutritional intake, and coastal populations use red seaweed regularly in sauces or salads.

In the 1970s and 1980s, the south of the country – where production is concentrated – was plagued by violent conflicts with Muslim minorities. Many Filipinos had to flee the violence with their families on small makeshift boats and migrate illegally to Sabah, one of the states of Malaysia. The so-called 'sea gypsies' still live by fishing and gathering from their boats, and took only a few seaweed cuttings from their native region.[125] Some are still working illegally in Malaysia today and have turned the country into one of the world's leading producers of seaweed.

Al-Jeria Abdul, a Malaysian phycologist, is descended from these migrants and tells their eventful story filled with anecdotes from a life spent on boats, going from island to island. Based in Sabah on the island of Borneo, she has an unwavering enthusiasm for seaweed. Al-Jeria is now the technical manager of a start-up[126] developing new genetic strains of seaweed that are more resistant to climate change.

Seaweed played an original and essential role in restoring peace in the Philippines. In 1996, after 25 years of civil war, the Moro National Liberation Front (MNLF) was finally considering signing peace agreements with the government, under the benevolent eye of the United States, for whom the Philippines remained a strategic area. The authorities soon realized that the 40,000 soldiers of the MNLF – who for twenty-five years had lived off looting and had no education to speak of – would face difficulties in finding employment in a region already ravaged by war. Moreover, the negotiations were moving towards a peace agreement, not a surrender. It was therefore unthinkable to claim the thousands of guns at their disposal. The combination of unemployment, poverty and the massive circulation of weapons left little room for dreams of serene and lasting harmony in this troubled region.

The CIA therefore decided, through USAID (United States Agency for International Development), to organize a vast exchange operation. They offered MNLF fighters the chance to hand in their weapons in return for training in algaculture and some ready-to-use cuttings.[127] And this is how more than 800 soldiers would trade their AK-47s for a handful of seaweed and the hope of becoming an ocean farmer... The former commander-in-chief of the MNLF was so successful in this new business that he even earned the nickname of the 'Seaweed King' in the region.

Poverty is not the sole cause of extremism, but the lack of economic prospects always contributes to exacerbating social violence. This new resource, seaweed, has demonstrated its power for social inclusion. For several years, however, the seaweed industry in the Philippines has been suffering. Conflicts, typhoons, disease and parasitic algae are all factors affecting productivity. Here too, the sector will have to reinvent itself in order to compete with its immense neighbour which has grown so quickly in its shadow: Indonesia.

With more than 13,000 islands, the largest archipelago in the world has no geopolitical conflicts or typhoons. Its 99,000 kilometres of coastline offer an enormous potential for cultivation. Indonesia represents 90% of the territorial waters of the Coral Triangle. Seaweed farming developed later there than it did in the Philippines and only really saw a boom in the early 2000s. The fourth most populous country in the world has now become the second-largest producer of seaweed in the world after China, with 30% of global production. The archipelago provides more than 80% of the seaweed currently used in the production of carrageenans. This raw material brings in around half a billion dollars each year.

The way Indonesian villages operate is unique and almost at odds with the capitalistic, patriarchal and productivist vision of much of the rest of the world. The organization of their culture is archaic and is based on the community life of the villages. Women are as well represented as men. There is no hierarchy or notion of private property on the farms. Certain tasks are designed to involve more family members in the collective work, not to achieve a better output. Often, the objective is to reach a certain level of monthly pay, but not to maximize revenue. This work philosophy creates paradoxical situations in which, when rates rise, the level of production falls.

The recent Covid-19 pandemic has further increased or renewed interest in seaweed farming as a vocation. In Bali, where money from tourism represented almost half of the island's income, everything disappeared overnight. Many guides, waiters, hoteliers, taxi drivers, diving instructors and others soon turned to seaweed farming as a source of income. Simple and requiring little investment, seaweed farming has become an ideal backup economy.

Yet it would be wrong to assume that the market for red seaweed for carrageenan is still flourishing. In fact, it is largely generated by Chinese demand, which is extremely volatile. And local communities have no financial reserves.

The lack of innovation and co-products puts producers and workers at the lowest level of the processing chain, in a state of exclusive dependence on a single end-product, the dynamics of which they do not control. The initial collaboration between production value chains and foreign buyers has largely deteriorated in recent years. Relations today seem to be limited to a violent price war.

Moreover, while demand is stable, the number of countries producing seaweed continues to increase, thus diluting investment capacities. In Indonesia, a company headed by Ian Neish – the son of Arthur, the early Canadian pioneer of red seaweed alongside Max Doty – is working to enhance the value of carrageenan as a biostimulant for plants. Using high-precision cultivation, Ian intends to take advantage of the active compounds extracted from *Kappaphycus* and *Eucheuma* while recovering a small amount of carrageenan in the process. He is also evaluating how to use the residues to make biodegradable packaging in a country with one of the highest rates of plastic pollution.

There needs to be many more efforts like this to allow the sector to renew itself and find a second wind by freeing itself from being dominated by foreign manufacturers. If Indonesia can rise to this challenge, it could lift a few more million people out of poverty and confirm its status as a major power in the future.

China also plays a key role in the Coral Triangle. It accounts for most of the world's production of agar and alginates, other texturizers extracted from seaweed, and remains the world's largest purchaser of semi-refined carrageenans in these southern countries. Moreover, the ties arising from shared ethnic origins with actors in the Coral Triangle value chains allow China to impose ethical and environmental practices that are often far removed from Western requirements, if not simply considered illegal.

The two blocs in North Asia and the Coral Triangle grow 99% of the volumes they send to market. In the rest of the world, by contrast, 99% of seaweed comes from a wild resource that is beyond control. Due to lack of experience, yields for similar seaweed are sometimes ten times lower than in North Asian countries. This gives the latter a strong competitive advantage in terms of price.

Asia exports seaweed; the rest of the world imports a lot of it. Asia is managing the regeneration of its seaweed, while the rest of the world is depleting its stocks and could soon lose a large number of species due to excessive and uncontrolled harvesting. Asia accounts for more than 97% of world production; Europe 0.8%; France, the world's second-largest maritime territory, barely more than 0.1%; the United States and Canada combined, 0.05%...

Europe and America: developing regions

However, large industries have been built around seaweed in Europe. In Scotland, in the eighteenth century, almost 100,000 people worked to extract potash and soda from seaweed to make glass.[128]

In Brittany, seaweed harvesting has for centuries been an essential activity for the coastal populations. Originally, seaweed was used to enrich the soil. Brittany owes its agricultural development and its valuable vegetable crops to seaweed.[129] The rapid growth in demand caused by the discovery of iodine in 1811 changed the sector and gradually led to the appearance of dedicated boats and tools. Throughout the nineteenth and up to the mid-twentieth century, thousands of people made a living from this activity. The discovery of mines destroyed the industry, but Brittany remains very active on the subject of seaweed, benefiting from a biodiversity that is unique in the world since it has 700 species spread over 2,400 kilometres of coastline,[130] i.e. as many species as in Australia but over 25,000 kilometres (metropolitan France has around 850 species of seaweed in total). There are 124 companies there, working on marine biotechnology from production to processing. This economic cluster innovating in seaweed, which employs more than 2,000 people with a turnover of 450 million euros in this region alone,[131] is unquestionably one of the most dynamic in Europe. A few years ago, the young company Algolesko created, in the south of the region, one of the first large-scale seaweed farms on the continent thanks to an offshore concession of more than 250 hectares.

Although Europe missed out on the start of this development, all is not lost. The surge in funding in this sector in recent years in Europe is a clear sign. The amounts invested in seaweed increased twenty-four-fold between 2010 and 2020, from 0.9 to 22 million euros per year over the period, with a clear acceleration from 2010.[132]

Over the last twenty years, the number of companies created per year has doubled from twenty to forty, and the number of investors from a wide range of backgrounds continues to grow.

The main producing countries are, in order of importance, Norway, France and Ireland. It isn't surprising that Norway is one of the countries in Europe that invests the most in the sector. With 85,000 kilometres of coastline, according to some calculations, Norway has the second-largest coastline in the world (more than France, the UK, the USA and Australia combined).[133] These coasts are bathed in cold waters, where large brown seaweed thrives. Moreover, Norway is well aware that the revenue windfall from fossil fuel extraction will not last, and that its aquaculture, the way they practise it, is less and less compatible with the country's environmental commitments. It's time to look for alternative resources that will allow the country to remain among the richest in the world.

In Ireland, the tradition of seaweed is old, especially on the west coast. One twelfth-century poem describes a monk collecting dulse on the beach to give to the poor. This seaweed has since been widely used in this region for food, medicine and cosmetics.[134] In the eighteenth century, seaweed ash was used to make soap or glass, and dulse was even sold as both a worming agent and a means of calming the 'sexual urges of men'. The widespread use of seaweed reappeared in food in the nineteenth century during the great famines which resulted in a rather negative perception of seaweed as a food of the poor. The University of Galway is still a centre of excellence in this field and is recognized at the European level, while some fine seaweed companies still operate in Ireland.

But to build experience in growing these vegetables in the sea, you still need the right to operate there. The main obstacle for seaweed producers in Europe and the United States is identifying suitable areas for cultivation and, above all, obtaining concessions from the authorities. We will come back to this subject, but licences are rare and issuing them is immensely complex. The profitability of operators is, however, linked to the size of their operation, and even if this varies greatly according to the type of seaweed, below a certain surface area it is not possible for them to survive.

Ease of access to operate in certain areas is a crucial element in allowing the development of a genuine socio-economic fabric for seaweed in Europe.

In the United States and Canada, the situation is very similar in many respects. There are cultivable areas all along the Pacific and Atlantic coasts but, as in Europe, production is hampered by delays in obtaining licences and by food regulations that are both confusing and restrictive.

In the United States, the main production areas are Maine, with its history of seafood products, Hawaii and California. California now has to restore the immense marine forests that have disappeared. Maine has a very strong tradition of seafood consumption and already has a certain culture of seaweed consumption as well. Most of the seaweed produced in the United States today comes from the coasts of this state, where the first company producing and marketing kelp was created in the early 2010s (i.e. more than half a century after the Chinese...). In the 1980s, attempts were even made in Washington State and then Maine and Alaska to grow *nori* to meet the rapidly growing demand linked to the Japanese restaurant boom. Unfortunately, despite a very favourable environment for the cultivation of this seaweed, its development was halted both by complex health and food standards and by the difficulty of obtaining permits in these regions.[135] In the 2010s, the US government also launched a major research programme to develop seaweed as a source of biofuel.[136] The focus quickly shifted to food products after realizing the limitations of its applications for energy. This large-scale programme led by the Department of Energy has greatly boosted investments and the progress of seaweed cultivation in the United States.

Experiments are currently underway to grow seaweed in the heavily polluted Long Island Sound in New York, to take advantage of seaweed's properties to clean up these coastlines.

Bathed in cold waters, Canada has the longest coastline in the world at over 200,000 kilometres, not including islands. This length represents more than a sixth of all of the world's coastlines and is greater than the sum of all three countries with the most coastlines (Norway, Indonesia and Denmark-Greenland). This gives Canada exceptional potential, even if there are many difficulties exploiting this often inaccessible space. The country also retains some of the original pioneering companies that somehow survived the end of the *Chondrus* adventure when tropical red seaweed arrived in the middle of the last century. New large-scale projects based on the ancestral knowledge of the First Nations are being set up for cultivation and processing, particularly in British Columbia.

Not far from there, seaweed has been recognized as an essential aspect of economic development. The state of Alaska, for example, has an unemployment rate twice that of the rest of the country.[137] This situation is due to the decrease in fisheries and the slowdown in investment in fossil fuel extraction. The state is in serious crisis and only survives thanks to huge subsidies from the federal government. Yet with a strong culture of working at sea, seaweed offers a huge opportunity for growth. Alaska has more than half of the total United States coastline and local authorities have estimated that the seaweed industry could generate a business volume of almost a billion dollars within thirty years.[138] This figure represents half of the revenue from fishing in

Alaska, the historical foundation of the country's identity. With the help of seaweed, local coastal communities could move towards a more resilient model which relies on varied sources of income, operates in a regenerative way and provides a wide range of ecosystem services. One of the largest seaweed farms in the United States was started in Alaska by a former lawyer turned seaweed farmer, who plans to soon produce hundreds of tonnes of endemic kelp solely for the food market.

South America is not to be outdone, and sea vegetables are an increasingly important resource in a continent where half the population lives in rural areas and where food insecurity affected 50 million people in 2019.[139]

In Chile, the world's largest producer of wild seaweed, more than 30,000 people work in the sector, with women well represented. Ever since the inhabitants of the Monte Verde cave 14,000 years ago, the seaweed industry has been an institution there. The 750 types of endemic seaweed and the long coasts lined with giant kelp forests helped to feed the people, although the arrival of the conquistadors stamped out some of these traditions. The fact remains that *cochayuyo* (*Durvillaea antarctica*), cooked in a stew, is an excellent substitute for meat and remains a very widespread dish in the Mapuche and Quechua populations in southern Chile.

The country produces equal amounts of brown and red seaweed, some of which is used for direct consumption while some is processed for the production of biostimulants, food texturizers and animal feed. This sector has serious advantages because it is not in the hands of a few landowners, does not require expensive inputs and is less threatened by natural disasters (such as drought, storms, fires, earthquakes) than land-based crops. Finally, the strong presence of the salmon industry could provide hope for integrating the two sectors when the model is in place. The combination of these two products could make salmon farming more environmentally friendly, since seaweed would capture part of the effluents that come from the fish farms.

Alas, the sector in Chile is currently based on harvesting a resource which, even though it seems abundant and free, is not limitless. We are witnessing here what some call 'the tragedy of the commons'.

With the government providing ample incentives for local people in need to engage in the seaweed industry, there is the risk of over-exploitation. Even the gigantic *Macrocystis* forests are beginning to shrink. Besides the gradual disappearance of this source of income for populations in economic distress, these plants which have been around for millions of years are an exceptional ecological niche and the foundation of many ecosystems. Their disappearance could have tragic economic, social and environmental consequences.

Chile has exceptional geographical conditions, a great variety of endemic species, an economic fabric, traditions and renowned researchers in the field. It is the eighth-largest seaweed-producing country in the world behind seven Asian countries, and the only country outside Asia and East Africa that grows the red seaweed *Gracilaria* on a large scale for the production of agar. However, Chile must continue its transformation, to enter into modern and regenerative production based on the cultivation of its other endemic seaweed.

Africa and India, a sea of opportunities in the face of a demographic challenge

Of all the developing regions, two of them could represent the greatest potential both in terms of their demographic dynamics – which will represent a considerable challenge in terms of resources – and the coastal conditions available to them. The first is Africa, which is perhaps the most exciting challenge for seaweed.

The experience in Tanzania is unique and, to this day, an exception on the continent. It is also the only country outside the Coral Triangle to export large volumes of red seaweed for carrageenans, and its model seems to be attracting a following. Interest in seaweed is growing in Kenya, Mozambique, Mauritius and South Africa. In South Africa, seaweed farming has existed for a long time but is limited to the production of green seaweed to feed the abalones that are exported at high prices to China.

In Namibia, the enormous Kelp Blue project in Lüderitz, which we will discuss later, promises a change of scale for farming in the open sea and provides highly valuable jobs and resources in a desert region desperately in need of them. Central Africa and West Africa do not offer any large-scale visions, even though small-scale projects are developing there.

We must go as far up as the Maghreb to find significant seaweed industry. Before the growing of seaweed developed in other countries, Morocco was the world specialist in and the first supplier of *Gelidium*, a genus of thalloid red algae intended for manufacturing the precious agar-agar. Today, Morocco's seaweed only represents a small minority of agar production, and the kingdom has destroyed much of its resources by overexploiting them. The government has tried to impose quotas on harvests, but these have not been respected.[140] It is always hard for populations in need to give up a free resource that is in such high demand on the market. Here again, major ecosystem disruptions are feared, with the predicted disappearance of these species. Alongside COP 22 in Marrakesh in 2016, the kingdom launched the 'Blue Belt Initiative' aimed at the emergence of a high-performance, low-carbon fisheries economy throughout Africa. Unfortunately, the operational progress towards the world of seaweed farming is currently still minimal.

Close by, Tunisia produces little seaweed but can boast one of the pioneers of the sector in Africa. In Bizerte, one of the first large-scale cultivated seaweed processing plants on the continent is now in operation. Supplied by Tanzania, Mozambique and local lagoon production – with the main aim of cleaning up stagnant water – the plant delivers refined products that are more profitable than the raw material. It sets an example for the entire continent.

Finally, the most inspiring and innovative model in the whole African continent may well be in Madagascar. There, after a promising start to cultivation initiated by the heirs of Max Doty in the 1990s, the sector was almost reduced to nothing by diseases and parasitic algae in the early 2010s. A few survivors then embarked on an innovative approach that drew on lessons from the past. The project started in the Toliara province in the south of the island. The local authorities, seeing the dangerous decline in the numbers of fish, decided to ban fishing for a quarter of the year to give the fish stock time to recover. This regulation put fishermen in an even more precarious and unsustainable situation. The government, with the help of European development funds, launched a project in partnership with NGOs and international entrepreneurs to promote seaweed farming. The initiative led to the creation of a company, Ocean Farmers, which was able to set up a new model for village seaweed farming in Africa. This operation was carried out with the support of one of the largest carrageenan manufacturers in the world, no doubt anxious to protect its image and to perpetuate a sector in danger by avoiding competition from the South East Asian markets.

Ocean Farmers has already provided seaweed farming training to more than 2,500 people from the surrounding villages. A contract has been signed with the seaweed farmers, who are generally fishermen or migrants from the interior who have come to the coast to escape poverty. This contract provides them with free technical support, a complete range of equipment, solutions to make their working methods more sustainable and, above all, a regular purchasing service at a guaranteed price. The farmers, in return, agree not to abandon nets or plastics at sea, to use only equipment provided by the company so as to avoid taking wood from the mangroves, and to respect the cultivation areas so as not to damage the seagrass beds or corals.

The fishermen work as a family, and meetings are held in village assemblies within each community at least twice a quarter. These island seaweed farmers, who have succeeded in domesticating the more profitable *Kappaphycus* species, have opened up a market with higher added value and can contribute to the survival of the model. This species, which can be grown all year round and which reaches maturity in forty-five days, allows Malagasy producers to benefit from a regular income thanks to a system of rotating cycles in the fields.

The results of this experiment are for the moment very convincing and the World Bank recently started supporting the initiative. The company is now expanding into other regions, including Nosy Bohara (or Sainte-Marie), off Madagascar's east coast. There, Ocean Farmers is experimenting with co-producing sea cucumbers, a very popular dish in China, where it will then be exported. These invertebrates will be fed by seaweed residues, which will avoid creating imbalances in the marine sediments. Although production is still in its early stages, a new, ethical and efficient model of village seaweed farming seems to be taking shape in one of the poorest countries in the world.

The second region that is expected to develop strongly in the coming years is the Indian subcontinent. India, despite having 800 endemic species of seaweed, has never really considered this resource other than for traditional medicine. This country will soon be the most populous country in the world, its lands are depleted by years of intensive agriculture and its land space is saturated both by urban sprawl and by the demand for food production. Almost 200 million people there suffer from hunger and 43% of children suffer from a state of chronic nutritional deficiency. Despite this, this country of nearly 1.5 billion people, with 7,500 km of coastline, still produces half as much seaweed as the million inhabitants of the microscopic island of Zanzibar in the same Indian Ocean...

In 2006, however, PepsiCo financed in the south of the country, just across from Sri Lanka, the first large-scale production of *Kappaphycus*. The US multinational hoped to show its support for the country's development as well as to create a new source of supply to meet its needs and those of its client, Mars, Inc. After five years of operation, PepsiCo's facilities were accused by local NGOs of damaging the corals, and the US company chose to withdraw to avoid litigation on such a sensitive issue. The local buyers, denying any damage to the coral ecosystems, continued their activity, somehow surviving a tsunami that destroyed all production in 2017. The company now employs more than 600 people, mostly women, and remains the only real structure for seaweed cultivation in India. PepsiCo's experience shows the difficulty faced by private investors when trying to finance innovative projects in the absence of international standards and regulations defined on a scientific basis.

In 2020, the Indian government announced funding of around 80 million dollars to promote the development of seaweed farming, marking a step in the right direction.

In neighbouring Bangladesh, the government, which had banned fishing since 2017 on the Myanmar border to allow fish stocks to replenish, announced in 2020 the launch of a plan to train former fishermen how to grow seaweed.

South Asia, Pacific and elsewhere: seaweed around the world

Thailand, whose inexperience in this field is more similar to India than to the neighbouring Coral Triangle, has nevertheless had a great seaweed success story. In 2004, when there was no tradition of eating sea vegetables in this country, the young Thai video game fanatic Itthipat Peeradechapan invested his first earnings in a project making seaweed chips, which he would personally sell outside the office to businessmen in Bangkok. The young man was only nineteen years old, but the product, with its beneficial nutritional profile as a snack, was a huge success. Demand skyrocketed. Less than fifteen years on, his seaweed is sold in all its forms throughout Asia and the Pacific, his company is listed on the Stock Exchange of Thailand, and Itthipat was cited as one of Thailand's 50 Richest in 2018, with a fortune estimated, according to *Forbes*, at more than 600 million dollars. His incredible story has been adapted into a movie, *The Billionaire* (2011).

It is inspiring to know that this kind of story already exists for sea vegetables, and that, as long as the product is good, integrating it into people's dietary habits can be quick and extremely profitable.

A little further south, in Australia and New Zealand, there is the same potential for seaweed. There is no shortage of coastlines and the conditions are ripe for it. Both countries have a wide variety of climates that favour the presence of a large number of species. The south of the area, especially Tasmania, is very rich and beautifully diverse, even if, there too, the immense kelp forests have disappeared from the combined action of climate change, deepwater dredging and invasions of sea urchins.

In the land of kangaroos, seaweed has historically aroused even less interest than anywhere else. Yet in 2019 a major national strategy, the Australian Seaweed Blueprint, was announced with the aim of developing the sector in order to reach 1.5 billion AUD (1 billion USD) in turnover by 2040,[141] generate 9,000 jobs and eliminate 10% of greenhouse gases along the way. In addition to the huge *Macrocystis*, the country benefits from the abundant presence of the famous *Asparagopsis* and its promise to suppress methane emissions in ruminants. This sector alone could well be a powerful source of economic growth for the country if this promise is confirmed over time by science.

A new redistribution

Fledgling projects exist elsewhere, from basin-scale farming in the Saudi Arabian desert to the frozen waters of the islands north-west of Russia. The melting of the ice should open up very interesting growing areas around the poles.

A young company from the Faroe Islands is already the European leader in the production of cultivated seaweed. In 2022, this company succeeded in bringing together the cream of the European seaweed industry and research into seaweed to obtain 9 million euros in funding from the European Union (Horizon Europe programme). This programme aims at the production and marketing of twelve different seaweed-based products.[142] The launch of this highly ambitious project saw all of Europe's leading seaweed producers come together in the tiny capital of the Faroe Islands in order to implement the project across the continent. As we enter the 'ocean civilization', the geopolitical balance of power may be shifting significantly.

This change will only be made possible by market demand. In the eighteenth century in Scotland, in the nineteenth century in Brittany and in the twentieth century in Canada, in the Coral Triangle and in China, Japan and Korea, seaweed industries have always developed to meet increased needs. It is therefore up to each of us to express this demand.

The seaweed industry also seems to be tempted, in some of its recent developments, to integrate production, processing and sometimes even distribution. This trend is much more marked than in the traditional agricultural sector. This change is also interesting. After all, it's clear that the system proposed by our current food organization is neither satisfactory nor fair. Most of the margin is captured at the processing and distribution stages; intermediaries add costs and reduce transparency. Meanwhile, farmers struggle to make a living from their work and survive only on state subsidies in rich countries, while those in poor countries starve. Remember that, according to the World Economic Forum, the vast majority of the 800 million people living in food insecurity are farmers, livestock breeders or fishermen.[143] As processing technologies become more accessible and distribution moves online, value chain integration at the producer level offers a new model that is worth observing.

Whatever the final outcome, this seaweed market already provides coastal populations with greater resilience. Because the economic outlook on these coasts is challenging. There is a combination of staggering population growth in regions already plagued by poverty, a largely degraded environment where more pollutants are pouring out every day, and a lack of economic prospects due to the depletion of local resources. Seaweed could contain some of the solutions. It creates new economic opportunities for these communities and opens up an unexplored field of innovation. It can also contribute to gender equality and women's empowerment through the creation of an inclusive and truly regenerative aquaculture model.

Above all, it offers us a way to reduce the growing inequalities that are a disgrace to our civilization and a ticking time bomb for all of us.

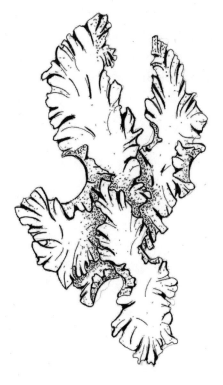

Porphyra (*nori*, or 'purple laver') – a very fine red seaweed known the world over for wrapping sushi, in the form of dried sheets. These sheets are made like paper, except that no artificial glue is added; it is the natural glue of the seaweed that binds the product. Growing up to 35 cm in length, it ranges in colour from light to very dark purple and its taste is often likened to mushroom or smoky tea, or even roasted hazelnut. Its high content of protein, omega-3, vitamin B12, iron and other nutrients make it very valuable, especially in Japan, where it used to be reserved for emperors, and where, eaten in various forms, it is central to the Japanese diet. There, the price of high-quality *nori* can be fifty times higher than standard *nori*. Different species of *Porphyra* are found all over the world in temperate waters. They are easily gathered at low tide.

7

HOW TO CULTIVATE THE OCEANS

*1950, Kathleen Mary Drew-Baker,
the 'Mother of the Sea' in Japan*

Without ever setting foot in Japan, or ever realizing the significance of her discovery, Kathleen Mary Drew-Baker revolutionized the world of seaweed that had existed for thousands of years in the Land of the Rising Sun, and enabled the emergence of a highly profitable industry that is now worth billions of dollars and employs tens of thousands of people. She achieved this feat by publishing an article in a scientific journal in 1949. She was unknown in her own country, but the impact of her discovery was so great in Japan that she is still called the 'Mother of the Sea' to this day and is celebrated on 14 April every year, a day when many devotees flock to the temple built in her honour.

Kathleen was born in England at the turn of the twentieth century. Her abilities as a young woman made her stand out: top of her class, and after graduating in botany from the University of Manchester, she became a lecturer in its cryptogamic botany department in 1922. After a fairly promising start to her career in the study of algae, she decided to get married and therefore, according to the academic rules at the time, had to leave her research position. At that time, married women did not have the right to be researchers at the university. But she was passionate about her subject and continued her research unpaid. This 'hobby' allowed her to produce almost fifty academic publications on red seaweed between 1924 and 1947.

She was particularly interested in *Porphyra umbilicalis*, or laver. This seaweed is harvested in Wales, where it is a popular base for soups and the traditional 'laver bread'.

A close cousin of this fragile little plant is widely consumed in Japan, where it is called *nori*. *Nori* is a staple of food in northern Asia, almost as important as bread is to the French. *Nori* is the seaweed which, once dried and transformed into sheets as thin as paper, is wrapped around the rice consumed daily by the population in this part of the world. This combination is called *sushi* or *maki* in Japan, and *kimbap* in Korea, and is a culinary tradition even more widespread there than sandwiches are in Europe. This dish has even developed internationally with dazzling success over the past thirty years, to the point of becoming the predominant symbol of Japanese food, although theirs is actually one of the most diverse cuisines in the world.

Nori is one of the most protein-rich seaweeds and is a key nutrient in the daily Japanese diet.

Harvesting *nori* has been an institution there for centuries. In ancient times, it is said that the shoguns (army generals) had bamboo barricades planted in the ocean to protect themselves from pirates. They found *nori* attached to the bamboo, when it was replaced. Fishermen then replicated the operation by multiplying the means to encourage it to grow. They had also noticed the presence of another microscopic alga, *Conchocelis*, which must have lived in symbiosis with the *nori* because it grew alongside it in alternate seasons. This tiny, very different alga gave a pinkish colour to the cockles on which it grew.

Due to the great variability of *nori* harvests and the difficulty of controlling its plant cycle, fishermen gradually abandoned this activity. The harvests were also so random that the seaweed was sometimes nicknamed 'lucky grass' or 'gamblers' grass'. As the prices of the raw material fluctuated greatly from year to year, *nori* was reserved for the wealthy classes.

Despite all their efforts and knowledge of the reproductive mechanism of *Porphyra*, the Japanese still could not overcome the hurdle of understanding the complete life cycle of this plant and enabling its cultivation. With no seeds, and no plants, there was a kind of 'missing link' in the biological history of *nori*: a mystery that no one could solve.

During World War II, this seaweed was solely consumed by army generals. Hence the need to produce a large amount and quickly break through its reproduction cycle. To do this, *nori* fishermen and researchers working on *nori* cultivation were exempted from combat. An unimaginable privilege when one considers that Japan requisitioned nearly 20 million soldiers, including women and fifteen-year-old children.[144]

But *nori* kept its secret.

For a long time, the instability of wild production had regularly caused social tensions in the country and plunged populations into poverty or famine. The situation worsened after the war, for reasons then unknown. This precious seaweed became increasingly rare in an already dramatic context: a country ruined and occupied by the Americans, a starving population, destroyed fishing boats and polluted seas.

The year 1947, which saw violent typhoons, was the worst year ever for the *nori* harvest. The population living in the traditional production zones sank into depression, which dealt a blow to the economy and undermined the country's food security.

At the same time, on the other side of the world in her Manchester laboratory, Kathleen was also studying *Porphyra* and inadvertently forgot to put away the shells she was studying in parallel. The consequences of this oversight were invaluable. Because it was then that Kathleen noticed that the shells had taken on the pinkish colour which indicates the presence of the alga *Conchocelis*. A strange idea came to her. What if the two algae were actually one? Like the plot of a dual-identity superhero movie where one of the secondary characters realizes he has never seen the two main characters in the same place, at the same time. *Porphyra* and *Conchocelis* were the Batman and Bruce Wayne of the ocean: they were one and the same!

More precisely, *Conchocelis* corresponded to a developmental stage in the reproductive cycle of *Porphyra*. Without realizing the importance of her discovery on the other side of the world, in 1949, Kathleen published a short article in the journal *Nature*, humbly putting forward this hypothesis. The consequences of this were unimaginable.

The Japanese phycologist Sokichi Segawa read these few lines in the journal and realized with amazement that *Conchocelis* was the missing piece of the puzzle, the one that would make it possible to grow this food which was so fundamental to the survival of the country's population. All the pieces then fit together as if by magic. *Conchocelis* develops in summer, *Porphyra* more so in winter. There are no plants or seeds, just spores. So production had dropped since the war because the shells in which they develop disappeared as the seabed had been damaged by pollution, bombardments and typhoons.

By 1953, based on Kathleen's premise, Segawa and his colleagues had already developed all the necessary techniques and spores for the planned cultivation of *nori*. The spores are delicately collected and the shells carefully selected. Production in Japan exploded. Thousands of Japanese fishermen were familiarizing themselves with these techniques and escaping from poverty. They regained their nutritional independence in proteins and above all, a little pride in their self-sufficiency.

With millions of tonnes produced each year and billions in profit, *nori* has become the most profitable aquaculture production worldwide. Its consumption has been exported from Bamako, to Rio, to Moscow. There isn't a single big city today where you can't eat sushi. It represents an unprecedented global culinary revolution! Such rapid success across all age groups is an exception in the usually slower field of dietary evolution. Nori and sushi have given Japan a new modern image and, together with manga and video games, have fostered the much-loved Japanese culture that has put the country back on the map.

Kathleen died young, in 1957, practically unaware of the effects of her discovery in Japan and all around the world. Six years later, the Japanese, forever grateful for this discovery, erected a temple in her honour at Uto, declared 14 April her celebration day and named her 'Mother of the Sea'. This title holds extra meaning in Japan, where the calligraphy of the word 'sea' already means 'Mother of all Japanese'. Kathleen thus became the grandmother of all the Japanese.

But Kathleen is not an isolated phenomenon. There are many incredible stories among these scientific pioneers who set out in the twentieth century, sometimes alone, to discover seaweed. Another example is Cheng Kui Tseng, in China.[145] This Chinese researcher, born in 1909, is recognized as one of the greatest seaweed specialists, and his life reads like a novel. Born into a peasant family, deeply affected as a child by the overthrow of the Qing dynasty and the invasion of Japan a little later, he was one of the first people to study seaweed. His work is widely praised. He was sent to do research in the United States during World War II, where he worked and shared his knowledge with Max Doty and his team. When he returned to China after the war, he continued his work on seaweed very successfully, despite very limited resources, and trained a large number of phycologists. Thanks to him, China is able to cultivate Japanese *kombu*, which is rich in iodine and other nutrients, thus helping to remedy the nutritional deficiencies of the population. Very early on, Tseng performed genetic crosses to enrich the existing varieties. He thus discovered over a hundred new species of seaweed and detailed their growth.

But in the mid-1950s the tide turned, with the 'Great Leap Forward' that equated scientists with the bourgeoisie. Persecuted and humiliated by the Red Guards, our seaweed genius went from biology researcher to cleaner at Qingdao University. While he was made to clean the toilets, he was also prevented from wearing his glasses. He was then imprisoned and tortured for years because he was considered an intellectual in the service of the Americans. In the 're-education' camp, he befriended close relatives of Deng Xiaoping and, through his courage and patience, managed to regain his place within the university, eventually becoming director of the Oceanographic Institute of Qingdao, then a member of parliament for the Qingdao region.

In the 1970s, he was one of the first scientists to be sent on an official visit to the United States to meet President Gerald Ford. There he met up with colleagues from his younger days, whom he'd lost contact with for almost twenty-five years. He was very impressed by the first computers and the progress of American technology, but was surprised by the poor progress made in phycology research. He had already contributed to developing this field of study and large-scale cultivation in a country that was initially less inclined than its Japanese neighbour to eat it. His tenacity and hard work made it possible to partly meet the nutritional and health needs of the population and to ensure its independence in terms of fertilizers and other chemical products. When he died in 2005, aged 96, near Qingdao, his country had several thousand hectares of seaweed fields – almost 60% of global production – which provided employment for hundreds of thousands of people.

This story shows how recent the field of phycology is. Without this knowledge, no profitable large-scale production would have been possible without risking over-exploitation of the wild resource, and this food would have remained a niche market for a wealthy population. Kathleen Drew-Baker and Cheng Kui Tseng were pioneers. There are still many discoveries and beautiful stories to be made. Access to knowledge about the genetics and behaviour of seaweed is a prerequisite for us to enter a new, more sustainable era, one in which both terrestrial and marine biotopes are well managed.

A civilization of the oceans.

How to grow seaweed

Domesticating, planting and growing seaweed by providing it with an optimal environment while ensuring that it does not damage the environment: this is undoubtedly the greatest challenge for this revolution. Our mastery of agriculture is the result of thousands of years of experiments, failures and genetic selection. It took a great deal of perseverance, attempts, errors and above all patience to acquire the necessary knowledge. Unfortunately, the urgency of the current situation means we need to move forward much more quickly with seaweed farming. The seaweed revolution requires significant capital and technology in research and industry, large-scale testing and, above all, working together to accelerate innovation and mastery of the production and transformation processes.

The practices for cultivating seaweed in the world today consist broadly of propagating seaweed according to two main modes of reproduction. In some species of red and green seaweeds, reproduction takes place by cuttings (also called vegetative reproduction) and part of the seaweed will create another identical seaweed. But some red seaweeds like *nori*, and brown seaweeds like kelp, reproduce sexually. In this case, male and female gametophytes emit gametes which are released into the water where they meet to form, after fertilization, the embryo of a new seaweed. These young shoots are then sown on nets or long ropes on the ground, then kept in the nursery for a few weeks to ensure their growth, and possibly increase the sowing density. Finally, they are transferred to farms at sea. In some species, the male gametes have two flagella instead of one, but their function is equivalent to that of mammalian spermatozoa – ensuring fertilization of the female gamete. This is another thing that seaweed and humans have in common.

Seaweed cultivation operates in 3D. Access to light is key, but the conditions of access are less binary than it may seem. On the one hand, the presence of seaweed will inevitably create shadow areas in the ocean. On the other hand, by absorbing the nutrients that float on the surface, marine plants reduce the turbidity of the water and facilitate the diffusion of light. This possibility of multi-storey cultivation, or 'vertical farming', opens up enormous potential for biomass production, compared to the horizontal cultivation we practice on land. We should remember that this crop, which is described as 'restorative' because of its many benefits, does not require any fresh water, arable land or pesticides.

Outside of Asia, most of the world remains in the prehistoric age of gathering wild seaweed resources. Here, simple scissors, sickles or 'scoubidous' (a kind of curved hook used for sixty years by seaweed harvesters in Brittany) have replaced the flints…

It has taken 12,000 years for that to happen.

Where to grow seaweed

Seaweed grows in salt water, or in fresh water. However, the pollution levels of lakes and rivers, their limited surface area and their already saturated use do not suggest great potential when it comes to fresh water.

So that leaves the ocean.

The ocean covers 361 million square kilometres of our planet. However, it would never be possible to cultivate the entire area. Seaweed, generally, needs rocky surfaces to attach to, unless artificial substrates are used. It needs a certain temperature level and must also be located at depths that allow it to capture sunlight, i.e. at a maximum of about 20 metres deep, although some species can develop as deep as 100 metres in very clear waters. Seaweed also needs to be in water that is sufficiently rich in nutrients for it to grow and with sufficient stirring to avoid accumulations of potential toxic substances (heavy metals, pollutants, etc.) which might make them unsuitable for direct human consumption. In theory, and taking only ecological constraints into account, the University of Santa Barbara estimates that seaweed cultivation is possible on a surface area of 48 million square kilometres.[146]

This gives us a bit of leeway if we consider that seaweed cultivation only currently covers less than 2,000 square kilometres of our planet![147] Today we use 0.004% of the theoretical surface area available for seaweed production. The figure is striking, but the living world is often much more than a simple equation. Indeed, it would first be necessary to recover enough nutrients from the land and to manage the impacts of algal residues in the marine sediment. There are also a number of logistical challenges which, while not insurmountable, are still very significant. Moreover, the most suitable natural spaces are coastal. They are generally characterized by a depth of less than twenty metres and together account for over 15 million square kilometres. Nothing can be done without these areas and the people who live there. However, these spaces are often unavailable because they are already used for maritime transport, fishing, tourism or even protected for the production of renewable energy at sea. Not to mention that a significant part of this area is already occupied by wild seaweed, which it would be unwise to eliminate. Competition is fierce for the ideal ocean location!

In addition, the coastal populations of these areas are still sometimes traumatized by destructive tides of seaweed, intensive salmon aquaculture or, for a minority, the detrimental effect on their holiday resort area. Added to this is the reluctance of the authorities to issue a permit or concession for an ocean area that does not belong to anyone, in order to carry out an activity there that is still little known and governed by rather confusing environmental regulations. Jeff Bezos, through his charitable fund and via the WWF, has recently allocated a budget of almost 100 million dollars to promoting projects that aim to increase the local acceptance of the cultivation of seaweed and seagrass by the populations.[148]

In order to overcome the complexities caused by existing maritime activities in these areas and the difficulty of obtaining exclusive concessions to grow seaweed there, new possibilities could still be explored. For example, 'sharing space' with other activities offers many opportunities. Fish farming, oyster farming, mussel farming or shellfish farming are promising areas for developing seaweed fields, pooling costs and creating positive exchanges between highly complementary farms.

Offshore aquaculture

If the coasts – being saturated by other activities or too polluted – do not necessarily favour large-scale development, then it seems logical to turn to the open sea. These distant, colder waters also seem to be the only way for the sector to become profitable, by combining high volumes and potentially large economies of scale. Unfortunately, as already discussed, surface nutrients become scarce as you move away from the coast. In order to increase cultivable areas in deeper areas of the ocean, the Japanese have devised tangles of ropes sown with seaweed that would descend to the bottom of the ocean at night to recover nutrients, then would rise to the surface during the day to benefit from the light. The cost still seems, in the current state of technology, to be prohibitive.

Seaweed co-production projects in areas dedicated to the production of renewable, solar or wind energy are being developed in the North Sea, thanks to funding from the European Union. The Netherlands is innovative in this area. These two activities can be carried out jointly. Offshore wind farms are beginning to mature as demand for more sustainable energy increases. They are rarely located in deep water, are protected from any passing boats, already benefit from concessions and offer good anchoring points for ropes sown with seaweed. Furthermore, this dual activity will make it possible to pool the maintenance, transport and repair costs. There are even already plans to add oyster and mussel production in order to develop the beginnings of integrated aquaculture, given that all these organisms already naturally colonize offshore wind turbines when the conditions are right. However, the solution is still a partial one, and cannot ensure development on a very large scale. The areas are limited to relatively shallow plateaus, the technology becomes more complex with regard to floating wind turbines, and access to nutrients often fluctuates in these areas.

As for nutrients, it is possible to recover them using natural methods in the open sea. One of these consists of using what are known as 'upwelling' zones: due to a combined effect of ocean winds and the earth's rotation, these unique and sometimes vast maritime spaces have very cold depths and a warmer water temperature towards the surface. This difference creates an upward current that allows the cold, nutrient-rich water to rise naturally from the abyssal depths towards the fishing zone. These conditions promote significant biodiversity, especially in phytoplankton and fish, but quite rarely in seaweed due to the lack of substrates for them to cling to. These currents are widely exploited because of the abundance of fish, and overfishing is already jeopardizing the resource's survival. Shifts in ecosystems have also occurred in one of the largest upwelling zones in the world, where gelatinous plankton (jellyfish) have supplanted fish.[149]

This upwelling zone, called the Benguela Current, in southern Africa, stretches from South Africa to Angola over approximately 70,000 hectares. Paradoxically, a vast, fertile water meadow stretches out opposite the Namibian desert, one of the hottest and driest deserts in the world, where local people are desperately short of water and food. Apart from purely extractive activities such as fishing and fossil fuel production, this huge area remains completely unused.

A recent project, Kelp Blue,[150] aims to take advantage of oil extraction experiments on the high seas to immerse cables sown with *Macrocystis* twenty kilometres deep, ten kilometres off the coast. *Macrocystis* is the incredible seaweed that has formed the bulk of the immense kelp forests of the Pacific coasts. It also occurs naturally in the Cape region of South Africa. As we have seen, this seaweed can grow a metre in a few days, reaching up to sixty metres in height and creating dense vegetation that is home to many species.

The idea is simple. As soon as the 'super-seaweed' reaches a few dozen metres, its canopy is harvested and then naturally dried out on the beach to then be recycled into various ingredients. It can produce food supplements for humans and animals, stimulate plants or provide natural textiles. This harvest takes place every three months and throughout the year. There is no need to resow it regularly, as this type of seaweed regrows well for at least seven years. After harvesting, the resource is dried on the beach without the use of any energy source, in a country where it is generally hot all year round and rainfall is scarce. And so, this giant kelp retains all its nutrients while remaining very lightweight, which allows it to be transported without a cold chain for many months across the continent.

Under these conditions, the prodigious seaweed absorbs a significant amount of carbon to create a biomass. As we have seen, it loses some of this as it grows. As the kelp is already located above a deep zone, much of this lost carbon is retained for centuries, millennia or even millions of years, in the seabed. According to Kelp Blue's calculations, if the optimum potential of the farm is cultivated, the amount of carbon dioxide absorbed will be equivalent to the amount emitted by a country such as the Netherlands. Given the international carbon offsetting mechanisms that should be developed for seaweed in the coming years, the company hopes to be able to achieve profitability even before it sells any of its production.

As already observed in all the kelp forests and adjacent areas, the presence of this culture will also allow the creation of a new ecosystem very rich in fish. This will reduce the risk of overfishing, promote biodiversity and sustain the incomes of local fishermen. The solution seems almost miraculous, even if there is still a long way to go in terms of mastering this new type of aquaculture. The logistical constraints of growing on the high seas – which are subject to violent waves and high winds and are where the largest sea animals roam – remain a real challenge for both production and transport.

Seaweed cultivation on land

Even if upwelling zones in the high seas or sharing space with renewable energies seem like attractive options, they will probably not be sufficient to realize the potential of our sea vegetables. We therefore need to consider combining complementary solutions. In certain circumstances, for example, seaweed farming in basins on land could prove to be advantageous.

The idea is not as far-fetched as it seems. The huge advantage of this system is that it provides complete control over the external parameters. Water, nutrients, temperature... everything is controlled. Of course, the production cost is often high and the logistics are complex. It is necessary to transport the water and the seeds and to fit out land that meets a certain number of criteria. But this operation is ideal for learning to master a species and for understanding its development in the minutest detail with a view to reproducing it on a large scale at sea. It is therefore very suitable for research phases on any scale.

Moreover, with this production system, the quantity is perfectly regulated. No whims of nature, swells or boats tearing cables; no shortage of nutrients or attacks by gastropods and other bacteria. The yield will also be high quality if the means are put into it. For high-value-added applications such as pharmaceuticals or cosmetics, this type of cultivation is particularly suitable. It will obviously never be possible to obtain large volumes with it, but the technique makes it possible to produce seaweed with perfectly calibrated properties for a precise and highly profitable use.

A number of projects of this kind are being developed in northern Europe. It's clear that the buyers of L'Oréal, who use a few grams of seaweed extracts in high-end creams with high added value, will not be as price-conscious as local grocery stores, but will be especially sensitive to the stable supply of high-quality material. The same will be true for drug manufacturers.

In another field, the previously mentioned company Acadian Seaplants,[151] has developed high-precision farming in Canada that is only possible in a closed environment on land for biological reasons. In its tanks it produces a very expensive but high-quality *Chondrus crispus*, which, above all, thanks to a process of selecting very targeted pigments, has bright and diverse colours. The algae are pink, yellow, green, etc. The Japanese adore them and are willing to pay up to a hundred times the price of common *Chondrus* to decorate their salads with these exceptional varieties. This niche production is only exported to Japan but remains profitable.

This basin-scale cultivation can also be adapted for desert or even arid areas. Middle Eastern countries, whose seas are too poor in nutrients to grow seaweed, are testing closed-circuit basins to enhance their deserts with algae that are light-hungry and flourish in high temperatures. These systems are often reserved for more profitable and productive microalgae, but trials are being developed with macroalgae.

Growing seaweed in basins on land is worthy of interest in very special circumstances but will probably remain quite a marginal activity. Intermediate and largely unexploited solutions still need to be considered: salt marshes, lagoons, estuaries or fjords. This type of cultivation between sea and land means the benefits of both environments can be enjoyed. The development of aquaculture will be a complex process which will require collective work and a capacity for innovation.

Domesticating seaweed

We need to continue to improve and deepen our knowledge and experience of phycology. Unfortunately, training experienced researchers takes time. So far, we have only domesticated between ten and twenty types of seaweed out of the 12,000 that exist. Only 200 species are on the market, most of which are wild seaweeds. According to the FAO, 95% of cultivated production comes from five different species of seaweed, mostly Asian.

Even the famous *nori*, so high in protein and so lucrative, has not been successfully domesticated in Europe. Only its Asian cousin can be cultivated thanks to the discoveries of Kathleen and her successors over the past seventy years. Our endemic *nori* still refuses to reveal all its secrets. Despite research and progress, European *nori* production remains at one tonne per year from a single farm in Portugal, which is still working to improve yields and hopefully produce more.

The other seaweed rich in protein and in high demand is dulse (or *Palmaria palmata*), a red seaweed found in abundance on the coast of Brittany and northern Europe. It is appreciated for its bacon-like flavour when grilled. It has experienced the same failure, as attempts at domestication are inconclusive and the wild resource is overexploited. Interesting developments and attempts to domesticate one of its varieties are underway, notably in Oregon, USA. For the time being, the supply is struggling to meet the ever-increasing demand, which also threatens the regeneration of stocks.

Seaweeds have complex and very different modes of reproduction. During 2022, it was even discovered that certain crustaceans help to pollinate red seaweed by travelling from males to females, like bees with plants on land.[152] This type of red seaweed has been on earth for over a billion years. So if its reproduction has always been aided by crustaceans, this dates the origin of animal pollination to much earlier than currently estimated, as well as the time when marine plants began to colonize dry land. Discoveries in this area are only just getting started.

While inspiration can be drawn from the work of Asian researchers, it is unfortunately not possible to replicate these models in Europe or America. It is therefore necessary to catalogue the types of seaweed present on our coasts and to domesticate them while controlling their proliferation, in order to avoid disturbances in the ecosystem. Investment and research are still nascent in Western countries in this sector. In a recent discussion with the director of Ifremer, the former director of INRA in France mentioned that there were 550 researchers in his country specializing in two relatively similar wheat species. A stark contrast to the seventy or so researchers working on 12,000 very different species of seaweed.

Understanding the language of seaweed

Among the most important recent discoveries, phycologists have demonstrated, for example, that seaweed has developed infinite levels of complexity so that they evolve with their environment, blending in and even communicating with each other! This communication is made possible thanks to the holobiont (or supraorganism). This is the assemblage made up of a natural, or multicellular entity (algae, plants or animals) and the cohort of microorganisms associated with it (now called microbiota). A seaweed thus forms a whole with the layers of microbes, bacteria, viruses and fungi that surround it. During its evolution, it learned to modulate these organisms to react to aggression, alert its fellow creatures to danger and maintain a healthy population.[153]

The components of these holobionts communicate via bacteria through what are called autoinducers. In this way, seaweed, like all marine organisms, perceives and emits chemical signals. For example, sexually reproducing brown algae release gametes into the water. The female gametes emit a pheromone, i.e. a sexual hormone (from the Greek *hormao*, meaning to excite, or arouse to action), which the male gametes detect and towards which they start to swim using their two flagella, orienting themselves according to the concentration rate of these substances. The emitters of these hormones exist only in female gametes and the receptors exist only in males. With seaweed, it's the girls who lead the dance!

These mechanisms also prevent inter-species reproduction of algae since it is also thanks to these substances that a spermatozoon from alga 'X' will never meet a female gamete from alga 'Y'.

Seaweeds are also able to alert each other. They do this in much the same way as the acacia tree, which when being nibbled on by a giraffe warns its fellow creatures by emitting a gas into the air. This gas triggers the production of a substance that gives their leaves a bad taste. When grazed by a periwinkle, *Fucus* also produces compounds that it releases into the water to alert its community, which perhaps uses the properties of the holobiont to create substances in their cells that cannot be digested by their tormentor. The reaction is much slower than for an acacia and the process takes several days to set up, but it has to be said that the periwinkle is not a particularly fast eater either.

When that is not enough, seaweeds even know how to call for help. *Ascophyllum nodosum*, which is widespread in the North Atlantic, has been shown to release odours into the water (and presumably into the air at low tide) when being grazed on by herbivores, thereby attracting herbivore predators such as fish and crabs. Other seaweeds are able to invite the natural parasites of their attacker. The molecules released have not yet been identified but seem to be emitted specifically when in contact with the saliva of certain herbivorous molluscs and not when the seaweed is damaged by mechanical abrasion or grazed by crustaceans, against which it will activate other defences.[154]

Most of these molecules and their receptors are still poorly understood. Deciphering the genomes of marine organisms is still in its infancy and we are unable to immediately decode this chemical language.

Yet it is this complexity that we need to decipher in order to learn to play in their key. A symphony orchestra the size of an ocean needs to be assembled.

Algal genetics and the pitfalls of standardization

The ambitions of the sector need to stay focused primarily on not falling into the industrial excesses that agriculture has experienced for several decades, with the establishment of large-scale ultra-mechanized and chemical monocultures that rely on a handful of highly productive species. Genetic diversity is essential. There is a great temptation to lean towards super-species or GMOs because it isn't possible to use pesticides or other plant protection products against external aggressors in the ocean. Many pioneers of seaweed farming have seen their production decimated in a matter of days by the sudden proliferation of marine grazers (gastropods) or by a sudden efflorescence of competing plankton.

Giving in to short-term convenience will only repeat the mistakes made on land. To illustrate this fear, we could cite recent research in Pennsylvania relating to the nine million Holstein cows.[155] These ruminants are believed to be 99% descended from two super-breeding males born in the 1960s. Two bulls with exceptional genetic potential are the sole male ancestors of these cows, which are therefore all cousins. Almost nine million cows, with only two different types of Y chromosomes... This ultra-efficient model has benefited the consumer, who has seen food prices fall, and also the breeders, who have achieved economies of scale. And it even reduced the environmental impact: fewer cows were needed to produce the same amount, and a cow's digestive system generates considerable amounts of methane. But those of us who remember our biology lessons in school will know that genetic homogeneity is never without long-term consequences. It increases the risk of hereditary diseases and, above all, reduces the resilience of a species by decreasing its ability to adapt to environmental changes. If a new disease breaks out, it will devastate entire herds of cows that share the same genes. We have seen many examples of this in the past.

The same principles should guide how we think about seaweed in the ocean. A cloned 'super-seaweed' would be immensely fragile in a much more complex and less controlled environment. The survival of this 'super-seaweed' would probably be much shorter than the American super Holstein cow infused with hundreds of litres of antibiotics for decades. As the seaweed industry was just getting started, it was unfortunately no exception. Reproduction by cuttings makes cloning seaweed even easier. And the development of the sector, although limited, has until now been too rarely accompanied by research efforts and real long-term vision. Repeated cloning of good-yielding varieties has already led to a high degree of species uniformity and a serious loss of biodiversity. In recent years and even today, this fragility has sometimes led to the sudden destruction of sectors, leading to great disappointment and human tragedy in the highly vulnerable coastal populations of certain emerging countries.

Understanding the genetics and maintaining maximum biodiversity among the different species of seaweed is therefore the only possible solution to ensuring the sustainable development of the sector.

Inventing seaweed farming 2.0

In order to develop this fledgling industry around the world, we need the Kathleen Drew-Bakers of the future to step forward! Marine plants have existed for over a billion years and their diversity and complexity is unparalleled. Future research should help us not only to understand the mechanisms of reproduction and growth of seaweed but also to counteract and prevent external attacks.

The genetics of algae has already made great progress over the past two decades, particularly thanks to rapid DNA analysis technologies. More than fifty genomes have already been deciphered since 2010 for macroalgae, revolutionizing our knowledge of both their evolution and of what they are capable of producing. We must still use the genetic potential of algae in order to select the most efficient and resistant varieties and learn how to grow them. We will need to ensure that the diversity of strains is maintained and that symbiosis with floating elements in the water is optimized.

Whatever people might say, new technologies are not just intended to dumb us down but are also a tremendous field of hope for finding solutions much faster than ever before. Already, companies in Norway and elsewhere are embarking on commercial solutions that enable the electronic sensor monitoring of water in seaweed farms and of species growth. Artificial intelligence systems retrieve this data and will soon make it possible, thanks to machine learning, to better understand these plants and optimize their growing conditions. These tools will also allow us to identify the best locations in the ocean for different types of seaweed and to anticipate external aggression from parasites or pathogens. They will show us the best cultivation periods according to underwater growth rates.

If public and private investors join forces, the results will be even better. The current trends give us something to be positive about. Both the European Union and the British and American governments are constantly increasing investment in this sector and allocating funding to new projects. The Green Deal, the main framework project of the EU strategy, clearly mentions seaweed as a key sector of the future. Change is also happening in terms of private investors. Up until 2013, annual private investment in algal biotechnology in Europe never exceeded one million euros, sometimes dipping to less than 500,000 euros. There has been steady growth ever since then, reaching close to 17 million in 2020.

Time is of the essence, so working together is fundamental. The first imperative is therefore to come together and collaborate. It is the responsibility of this promising young sector to avoid duplicating work, to share ideas, innovations, funding, economic models and scientific advances with regard to the genetic understanding of an unknown and extremely diverse resource, and the use of the spaces in which it is grown. Can we achieve in a few years at sea what has taken us millennia on land?

Our society has stayed afloat during the pandemic thanks to research, massive investment by the state and the unfailingly supportive mobilization of a number of professions, first and foremost those involved in healthcare and research.

Why isn't there the same urgency to save the planet and, by extension, humanity? It's time to make the environment a top priority and to believe in our technological progress. The oceans are the first victims of our excesses, and yet, if science succeeds in better understanding them, they could provide us with the solutions we lack.

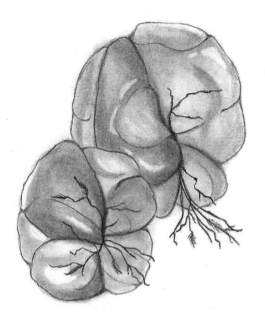

Colpomenia peregrina ('oyster thief') – a brown seaweed unintentionally introduced into French waters at the beginning of the twentieth century with the marketing of imported American oysters. It has developed spectacularly in French oyster beds and has now become part of the local flora of European coastlines. Its large seven-centimetre vesicles are filled with air, making it float. This is where it gets its nickname from, as it clings to oysters and carries them out of the oyster beds when the tide rises...

8

SEAWEED: ENVIRONMENTAL RISKS OR AN OPPORTUNITY TO SAVE THE PLANET?

*Christopher Columbus
and the Sargasso Sea*

On 7 October 1492, Christopher Columbus, aboard his ship the *Santa María*, had been at a complete standstill for twenty-one days, trapped in a very strange sea which he would later name the Sargasso Sea, 'sargazo' in Spanish meaning 'gulfweed'.

This sea is covered with it to such an extent that the ship, bogged down in this golden mass, seemed destined to stay there forever. There was great anxiety among the sailors on board as they saw the already heavily rationed food supplies dwindling. Not the slightest wind, not the slightest wave, only this tangle of seaweed as far as the eye could see.

Three weeks before, Columbus had already written in his notebooks: 'Here they began to see many tufts of grass that were very green. [156] After fearing shallows that would sink ships, the crew had hoped to see this as a sign that land was near, believing that this seaweed had been torn from the coasts and brought there by some storm.

But it was not. They could not know that the Sargasso Sea is six times the size of France and is the only sea in the world with no shoreline. And although land was near, the ship did not budge. In the sweltering humidity, panic mounted as there was no sign of wind on the horizon. This dangerous expedition would come to the strangest end, they all thought. That day, the crew was convinced that the *Santa María* had arrived at the very edge of the world, which was said to be round…

The end of the world was predicted to have howling monsters, hellfire, violent whirlpools or deep ravines, but it was a peaceful sea of grass that might ultimately serve as their tomb. Terrified by this prospect, some attempted a mutiny but eventually gave up. What choice did they have when the horizon was swamped with seaweed?

Columbus had already heard of this sea, however. A few months earlier, in a grimy bar in the port of Palos in Andalusia, a man had warned him about it. That man was Pedro Vázquez de la Frontera, a former sailor in the service of Portugal, who had been sailing the seas for decades. Vázquez was respected by all. Forty years earlier, as a young sailor, he had sailed with Diogo de Teive.

De Teive was a famous Portuguese navigator who sighted the two western islands of the Azores. During an expedition, Vázquez had followed Diogo even further west in search of a legendary island called Antillia.

A cartographer had confirmed, some years before Diogo and Vázquez's voyage, and long before that of Columbus, the presence of Antillia.

In any case, in Palos in 1492, Vázquez assured Columbus that Antillia did exist. He never found it, because they got bogged down in a thick, brownish mass that seemed to cover the sea. After two weeks of waiting, the wind picked up and they were able to turn back. He always regretted that decision… Vázquez advised Columbus not to be discouraged and to wait for a tailwind. The promised land must be there!

The idea is not new: two thousand years ago, in an ancient text by Persian navigators, it is written that 'one cannot sail beyond the island of Cerné (the westernmost island known at the time, its location remains a mystery), because the sea is embarrassed by mud and weeds'.[157]

Later, during the first century, the Carthaginians, long-distance travellers, told of these 'numerous algae below the waves which, by interweaving, form a thousand obstacles. Not a breath of wind pushes the ship forward. The waves are idle, motionless. Algae are sown in innumerable quantities over the depths, and often they stop the progress of the vessels which they hold back like rushes.'[158]

Later still, the Gothic historian Jornandes wrote that 'the reason why some regions beyond the oceans remain uncharted is that the seaweed stops the ships from moving and the winds have no strength'.[159]

Christopher Columbus was staunchly determined to be the first to force his way through and return. He would have to be patient, even if it meant losing his life and that of his crew. Columbus would wait, and not turn back at the first tailwind. He watched impassively as small crabs walked over the seaweed, seemingly taunting him. *Planes minutus*, now called the Columbus Crab, is one of the most adept at camouflage. Golden brown, it sometimes has a white shell and a blue belly. This protects it from birds that mistake it for a sargassum bulb and from marine predators that think they can see only the blue sky. Along with a whale, which Columbus notes in his diaries, these crabs were the only company for the immobile sailors trapped in their ships. Beneath the seaweed lurked thousands of eels and sharks.

A few decades later, the British called this area 'the horse latitudes' because ships, laden with the steeds of the Spanish conquistadors, had to resolve to eat the horses if the ocean held them captive for too long. This sea has given rise to many legends. Ships filled with dead bodies have been found there.

The long-awaited wind finally arrived. It was a light wind, but one that pushed them still further into that trap, which seemed to close in behind the ships.

Columbus very slowly sank into decline and the *Santa María* caravel allowed itself to be devoured by the sea of algae. However, a few days later, on 12 October, land finally appeared in the distance. Columbus and his men arrived on the shores of the Bahamas…

They had overcome the last great barrier that divided the earth into two distinct worlds.

A new era began!

Columbus was probably largely unaware of the reasons for the presence of all that seaweed. That wasn't his concern. Jules Verne goes further and, in *Twenty Thousand Leagues Under the Sea*, describes this strange sea: 'Such was the region [...] a genuine prairie, a tightly woven carpet of algae, gulfweed, and bladder wrack so dense and compact a craft's stempost couldn't tear through it without difficulty [...] a section of cool, tranquil, motionless ocean called the Sargasso Sea. This is an actual lake in the open Atlantic and the great current's waters take at least three years to circle it. Properly speaking, the Sargasso Sea covers every submerged part of Atlantis. Certain authors have even held that the many weeds strewn over this sea were torn loose from the prairies of that ancient continent.'[160]

Yet the reason for this incredible accumulation of algae over millions of square kilometres is not linked to the city submerged beneath the waves mentioned by Plato, but to a vortex created by the convergence of the three great marine currents of the Atlantic. An ocean gyre.

This body of water, with a motionless surface in the middle of the ocean, is quite remarkable. In the centre, the sea level is about a metre higher than on the east coast of the United States. The seaweeds found there are also very special. There are more than sixty different species of sargassum around the world. Some are found clinging to rocks in Europe as well as in Japan where they are a delicacy (*hiziki*). Two species of sargassum coexist in the Sargasso Sea (*Sargassum natans* and *Sargassum fluitans*).[161] These species are also found in the Gulf of Mexico, where some forms appear to have been abundantly fed for the past forty years by nutrients from coastal production and the mouth of the Mississippi River. Americans reported the first golden tides off the coasts of Texas and Florida in the 1980s.[162] In 2011, satellite images revealed huge quantities further south, at the mouth of the Amazon; from there, they would have started to proliferate until they invaded the entire Caribbean Sea.

Since 2011, the proliferation of these new sargassum in this area of the Caribbean has accelerated exponentially, multiplying its biomass tenfold in the space of a few years. The Great Atlantic Sargassum Belt now covers an area from

Central America to the coast of West Africa. From Texas to Abidjan, these huge rafts form a strip of 9,000 kilometres.

In the open sea, sargassum contributes to biodiversity by providing a refuge for turtles, crabs and fish. But when their proliferation takes them near the coasts at the bottom of the Gulf of Mexico, they prevent light from penetrating which causes the corals and seagrasses to die. Once dead, they decompose and absorb all the oxygen, creating deserted areas on the shoreline to the detriment of fish, crustaceans and turtles. On land, they cover the shores for several kilometres in heaps almost a metre thick, releasing a stench similar to rotten eggs as they decompose.

The hydrogen sulphide they emit causes eye irritation, nausea, sore throat or ears, or itchy skin. This phenomenon destroys all tourist activities in disadvantaged areas where this income is essential to the population.

The seaweed also clogs the propellers of fishing boats, preventing fishermen from working.

Today, these 'golden tides' are a scourge for the populations of Martinique, Guadeloupe, the southern United States, Mexico, the Bahamas and all of the Caribbean islands. Inevitably there are multiple reasons for this massive proliferation, one of which is overfishing of the main predatory molluscs that graze on these seaweeds[163] (although this cause appears to be very marginal). The real cause, most likely, has been identified through satellite observation of their evolution. In fact, these mountains of seaweed do not come from the Sargasso Sea but from a more southerly region, opposite Brazil, at the mouth of the Amazon.

The banks of the world's largest river have been suffering for years from the ravages of intensive agriculture and deforestation, and are spitting out an incalculable amount of nitrogen, phosphorus and other waste into the ocean. Exhausted and leached soils regurgitate hundreds of thousands of litres of pollutants and wastewater act on the seaweed like oil thrown onto a fire; everything goes up in an apocalyptic blaze. The seaweed proliferates and accumulates by millions of tonnes.

In 2011, there were 10 million tonnes of sargassum, in 2018 this figure reached 20 million tonnes.[164] A gigantic, high-speed development. By way of comparison, the tides of green seaweed in the north of Brittany amount, in their most significant years, to a volume of around 100,000 tonnes. In both cases, the cause is the uncontrolled pollution coming from the land surface, which fuels this propagation. Our civilization still seems to believe that this new regenerative agriculture can be confined to inside our shores. But this belief is absurd. Everything that happens on land ends up in the ocean. If we spill chaos, we will reap chaos...

The problem with the sargassum from the Caribbean Sea arises precisely from the fact that they are located in the open sea. This makes them even more difficult to eradicate, especially as their routes are almost impossible to predict. How can we control and develop something we can only recover once it has rotted on the beaches? Moreover, collecting these quantities of dead seaweed on land is an inefficient, heavy operation requiring significant manpower and financial and logistical resources. Studies are underway into offshore booms or boats specifically built to harvest them alive offshore, but the difficulty in predicting the location of the seaweed and the dates of its proliferation makes it complicated to implement these innovations. There are already some proposals to develop a means of sinking the seaweed in the open sea in order to sequester the carbon from its biomass in the marine sediments for thousands of years to come.

However, their compounds are also of interest. Sargassum cannot really be eaten directly because its levels of heavy metals and arsenic are very high. But if the proteins (8%) and carbohydrates (39%) were extracted from it, this would be very beneficial to populations suffering from major nutritional problems. It can also be converted into methane or composted. However, the most profitable use of this seaweed at present seems to be in its alginate, which can be processed into biomaterials. Local investors are demonstrating a growing interest in using sargassum in biodegradable packaging.

Other entrepreneurs are starting to manufacture building bricks using this sargassum. In 2018, a Mexican company built a house in two weeks from sargassum harvested on the coast of Puerto Morelos. It was dried and mixed with wet raw earth to form bricks which then hardened in the sun. According to studies by scientists from the National Autonomous University of Mexico (UNAM), a house made from these bricks can withstand earthquakes and strong winds. The bricks obtained by incorporating this brown seaweed provide much more resilience than traditional ones. This project aims to create more affordable housing for people on low incomes.[165] In late 2022, the local agency of the UN Food and Agriculture Organization (FAO) released a very comprehensive guide to all existing uses of sargassum in the Caribbean, listing its advantages and risks with the objective being to 'turn this hazard into an opportunity'.[166]

Industrial processing plants enabling these applications

would be a possibility, but again, working out where to put them is complex given the imprecise location of the algal bloom from year to year. Not to mention that we do not know how long this phenomenon, which we have only just begun to study, will last. Furthermore, investment capacity and the definition of regulations are also complex issues in a maritime space cohabited by fourteen countries, without any notable leadership.

The Sargasso Sea is still fed by these converging currents and now probably contains more plastic than seaweed. Yet if no solution is found, then the very sargassum that struck terror into Columbus's sailors could destroy the coastal resources – both biological and economic – of Central America and even Africa in the years to come.

Incidentally, media coverage of this scourge continues to unfairly tarnish the image of seaweed throughout the world. Images of sargassum rotting on beaches blur the message that we need to domesticate this seaweed, even though its harmful aspect is precisely due to its wild and uncontrolled nature. Have you ever heard of uncontrolled wheat proliferation?

In terms of reducing its current impact, there is unfortunately no indication that the countries bordering the Amazon are committed to taking action to limit deforestation, pollution and the influx of nutrients and minerals into the ocean. Because above all, we have to remember that sargassum is not the cause of the problem, it is the symptom!

But the historical symbolism remains. For a long time, this sargassum prevented the great navigators from moving forward to connect the two great evolutionary blocs of our world; the two civilizational models that had evolved in parallel up to that point. As soon as Christopher Columbus managed to tame the seaweed, our humanity tipped into a new era and the world became one.

What keys to progress might be hidden in seaweed and in our capacity to rediscover the ocean today?

The environmental risks and ecosystem services of seaweed

In environmental matters, seaweed is too often perceived as a risk, whereas, on the contrary, it represents an opportunity. Its bad reputation is linked to the economic and ecological damage caused throughout the world by tides of seaweed that we've been unable to control or fully understand.

These seaweed tides are generally only the consequence of human actions and the environmental disruptions caused by our agricultural and industrial system. Seaweed acts like the immune system of our planetary biotope when faced with an aggression: pollution by our society which, since the end of the nineteenth century, has opted for synthetic chemistry – often in defiance of the chemistry of living organisms – and has never integrated the ocean into its waste cycle. Chemical residues from fertilizers and animal excrements flow into rivers and slide over our soils worn down by intensive agriculture and, sooner or later, inevitably end up in the sea.

When a virus causes a fever, we can try to relieve the fever, but the most efficient thing is to treat the virus. Seaweed tides are not the problem; they are primarily a symptom. Addressing the cause of the problem will be a long process, as it involves major systemic change. As for the symptom, it is currently poorly managed because of our immaturity in all matters relating to the ocean.

In our conception of the world, the ocean is often outside our system politically, economically and often also ecologically. As the ocean belongs to no one, it is largely beyond the responsibility of individual countries. Its resources are shared or abandoned to an institutional vacuum. There is no UN agency responsible for the ocean; hence our inability to collaboratively prevent the major threats to it.

The ocean is seen as the edge of our nation states and usually represents our borders. However, it could well be a means of connection.

While Asia has turned seaweed into an ecological and economic asset, the West is in denial about the impacts of our terrestrial systems on our oceans, which we imagine to be still wild. At the heart of this problem is a lack of understanding of the mechanisms involved in the lifecycle of seaweed and our unwillingness to intervene and address the problem upstream or downstream. Northern Brittany's green seaweed is a perfect case study which in many ways echoes other red, brown, golden or green tides around the world.

Green tides: the consequence of land pollution[167]

After the Second World War, the public authorities in France, supported by the dominant American model, developed intensive agriculture in Brittany, which had productive land then. Seaweed had been widely used there for centuries to naturally fertilize fields and protect plants. At that time, extensive land-use planning reshaped the landscape. Huge swathes of hedgerows and trees were cut down and rivers were drained to create large plots of land that could be used for agricultural machinery. Many farmers and organizations were concerned about the consequences of such a change modifying drainage and reducing the filtration of the land. The disappearance of trout and salmon in the early 1970s soon proved them right, and the news made the headlines. But the world had already turned too far towards intensive agriculture to be halted in its tracks by a few media stories about fish.

This period also marked the end of a traditional type of breeding to make way for industrial breeding focused on a single species. Mono-livestock farming led to intensive, increasingly concentrated and indebted farms that were surviving on state support alone. The number of farmers dropped dramatically throughout France and especially in Brittany. The ports of the region, and Brest in particular, became key logistical centres for importing GM soya or maize and exporting animals to international markets. The pig farming industry, largely supported by the local and national authorities, became particularly significant.

In the 1970s and 1980s, the number of 'soil-less' pig farms grew at a rapid rate, and in some cases the expansions reached the limits of legality. Animals were crammed into pig factories and Brittany was forced to imitate the ultra-productivist models used in other countries. Initially, the development of the sector mainly served France. In 2016, Brittany still produced 58% of all French pigs on only 8% of the territory.[168] The Breton food industry had long been promoted as the jewel of French industry.

From the 1980s onwards, the consequences were felt in literally every sense. The water became undrinkable and a very strong smell of manure invaded the countryside. The use of slatted floors instead of straw is a good example of the excesses of a system that led to these green seaweed tides. The process originated in Denmark but was rapidly adopted by the Bretons. Until then, pigs were raised on straw. Mixed with pig slurry, this produced manure that was used to enrich the land, to promote the production of feed for the animals. The loop seemed efficient. However, the space required for these farms and the costs involved (purchasing straw, transport, spreading manure, etc.) were soon considered too high. And slatted floors – a system of wooden boards through which the pigs' droppings can pass – reduced costs further and intensified rearing in small areas.

This generated significant economies of scale. In the short term at least. Whereas manure used to preserve and enhance the pigs' slurry, slatted floors allow this nitrogen-rich substance to pass through, and once in the soil it is transformed into nitrate by the action of bacteria. The nitrates are then spread by runoff into rivers, which flow into the sea.

In addition to this practice, there is the pollution caused by widespread use of synthetic ammonium nitrate-phosphate fertilizers that farmers are obliged to use due to the lack of available manure to boost crops... The age-old balance between the soil, plants and animals was therefore broken.

It is essential to point out that these changes, which move us away from a system based on living organisms, were implemented above all to meet consumer demand for ever lower prices in an era of great technological transformation. This change was facilitated by an international food market that encouraged competition, while the rise of large-scale retail made it impossible for producers to negotiate alternative, less productivist models. It is crucial to acknowledge this. Today, it is more important to learn lessons for the future than to blame the ostensible perpetrators of the past. The shift was collective, on the part of the producers, authorities and big private brands. Moreover, the food market is so fragmented and globalized that trends are primarily defined by the daily choices of each and every one of us. And at that time, price was still practically the only factor that guided consumers' decisions.

To return to our Breton green seaweed: these industrial quantities of non-filtered nitrates and phosphates act as super-fertilizers. Their action is similar on land and at sea. The green seaweed, which was renamed *Ulva armoricana* in the 1990s, would therefore experience an unprecedented proliferation. The sheltered sandy bays of Brittany support the growth of this seaweed, which is characterized by a very fragile structure. It does not remain attached to its substrate for long but has a very rapid reproductive capacity during the summer. Areas in which the water is not stirred encourage the seaweed to rot, and it then releases hydrogen sulphide with a characteristic smell like rotten egg. The bacterial degradation of dead seaweed thus absorbs oxygen in waters that are said to be 'anoxic' (oxygen-less). As a result, all plant and animal life disappear. This saturation of nutrients in an area is called 'eutrophication' and causes serious disruptions.

The consequences are well known: these mountains of green seaweed destroy coastal ecosystems, tourism and continue to be a major health hazard for the population. In some bays the seaweed carried by the tides ends up decomposing in the muddy substrate to form dozens of cubic metres of toxic sludge. These are green, slimy deserts that reek of death. No insects or shellfish can survive. This putrefied seaweed has already killed a number of animals and even humans. However, there have been many efforts at the industrial and political level to hide this sad truth in order to protect a critical agricultural sector representing many jobs and interests.[169]

However, the situation is less complex than for the Caribbean sargassum, both due to the smaller volume and the coastal situation which facilitates live collection. The CEVA (Algae Technology and Innovation Centre) is now in charge of monitoring seaweed and working on innovation projects. At the same time, the Breton company Olmix, which is already well established for its seaweed-based products for animals and plants, has set up machines to harvest the seaweed at sea before it rots on land. Because this green seaweed, also known as 'sea salad', is an interesting ingredient when it is fresh – both in food and agriculture, but also as a food supplement because of its high content of vitamins C and B1, and antimicrobial substances. Some extracts even show interesting antiviral properties against Covid-19, as mentioned earlier. The strangest paradox is that *Ulva* is already used by various Breton processing companies which are sometimes forced to move to China or to develop green seaweed farming on land! In very small volumes, its local production in tanks is used to feed high-end aquaculture products such as abalone or can be used for pharmaceutical markets that require high quality.

Even today, the idea of harvesting this seaweed fresh in the bays before it dies and rots on land faces two major obstacles. The first is related to the use of machines that have an impact on the ecosystem of sandy beaches. Shellfish and fish would suffer from being crushed by equipment not designed to preserve this biotope. After years of consultation, a solution seems to have been found and suitable machines are now in operation.

The other obstacle is more of an ethical nature. It reflects the often political and militant refusal to extract value from resources that are the result of pollution and a toxic system. This reasoning seems legitimate and it is easy to understand that, if this nitrate pollution ends up generating income and jobs in Brittany, it will be all the more difficult to call for a transition from industrial agriculture to a more regenerative model. Unfortunately, doing nothing or simply storing or spreading the dead seaweed seems to be an even worse option.

2021 was a record year for green tides in Brittany, where the area of coverage exceeded the annual average since the turn of the century by 50%. However, the producers have made real efforts and over the last twenty-five years the average nitrate level in Breton surface waters has been reduced by almost a third. This improvement is still far from satisfactory, and the inertia of natural biological systems will require a lot of patience from us.

The problem is obviously not just a Breton one, but also a European one. In the Baltic Sea, unbridled agriculture in Poland is leading to even more eutrophication. For its part, in September 2021 the Netherlands announced a drastic plan to reform its food production systems to urgently reduce nitrate pollution of their waterways.[170] The Dutch authorities have always been innately committed to economic liberalism and are proud of their food industry, which is one of the most successful in Europe. The fact that they've decided to impose a form of large-scale nationalization and 'state' dismantling of the country's intensive farms speaks volumes about the urgency of the situation. Limiting the development of more sustainable agriculture to a purely terrestrial context is simply not adequate. The oceans must be integrated into the system to make it truly circular. Today, in Brittany, some people are advocating the development of models where seaweed would be used to clean up the oceans by capturing nitrates and phosphates.

In the bays of the Côtes-d'Armor region, at the mouths of rivers, we could imagine laying cables seeded with brown seaweed capable of capturing these pollutants before the green seaweed does. This brown seaweed is said to be nourished by agricultural pollution, which enhances its growth. Since it has a more robust physical structure than green seaweed, it would remain attached to the cables and, when harvested regularly, would not end up rotting on the beaches. It would then be sold and spread on the land, creating a truly regenerative and integrated agriculture. The green seaweed, on the other hand, would no longer have enough nutrients to proliferate and would return to acceptable reproduction levels.

How China is curbing its seaweed tides and recycling its pollutants

Similar techniques are widely used in China. For a long time now, seaweed tides have been recurrent and disproportionate. The most invasive seaweed on these shores is another form of *Ulva*, the aptly named *Enteromorpha prolifera*, which looks like green moss, and also goes by the name of 'sea lettuce'. One of these particularly virulent 'salad tides' occurred in Qingdao in the summer of 2008, when Beijing was hosting the Olympic Games. For the sailing events that were to take place there, more than 10,000 people were requisitioned to collect the 170,000 tonnes of seaweed. This slimy, smelly carpet probably did not fit with the image China wanted to project.

Five years later, the phenomenon became even more dramatic in Qingdao and, exceeding previous records, the Yellow Sea turned green over an area of 30,000 square kilometres. It must be said that the equation for enabling the management of land-based waste in China's coastal areas seems impossible to solve, at first glance. The coasts, with 14% of the land area, are home to almost half of the inhabitants of what remains the most populous country in the world.[171] On this strip of land, the average density is almost 500 inhabitants per square kilometre and can reach 3,800 in Shanghai, which is more than 35 times the French average population density. This situation leads to the discharge of wastewater that is rich in phosphates into the ocean.

Industry is also very present and ensures the production and exports of the largest and densest manufacturing area in the world. Seven of the ten largest ports in the world are Chinese. Throughout the country, intensive agriculture extensively uses organic fertilizers which invade the soil and enter the rivers, before then ending up in the ocean. Tibet is home to nine rivers that supply water to a quarter of the world's population.[172] More than two billion people dump their domestic, agricultural and industrial waste into these waterways.

All of these factors indicate that territorial waters will be clogged up with waste and these volumes seem impossible to manage. Fortunately, in China they are very familiar with growing seaweed and are aware that dousing seaweed with nitrogen can more than quadruple its growth.[173] And the more the seaweed grows, the more nitrogen – and therefore pollutant – it absorbs. So China recently decided to put its experience into practice by growing seaweed in order to restore the balance. This technique of recycling pollutants through seaweed, known as 'bioremediation', has so far proved to be impressively effective.

In 2016, international research was able to quantify the exact amounts of nutrients that these large-scale Chinese seaweed farms could absorb. The objective was to better understand its role in mitigating eutrophication in coastal areas. The experts concluded that on average one hectare of seaweed aquaculture absorbs the nitrates needed to treat 18 hectares of land crops. This figure rises to 127 hectares for phosphates.[174] For example, seaweed crops in China now absorb 75,000 tonnes of nitrates and 9,500 tonnes of phosphates per year!

Considering the growth rate of seaweed farming, it was estimated that by 2026, seaweed could absorb all the phosphates discharged into the ocean in China. These phosphates in the seaweed can then be reused to feed crops on land. This capacity is very valuable in terms of both preventing pollution and creating phosphate resources. Like oil, phosphate is now produced by mining, making it a non-renewable resource. According to all the specialists, these phosphate resources will soon be in short supply. This substance is a prerequisite for plant life on earth. In this context, the ability to recycle phosphate for reuse on land via seaweed becomes a truly strategic – even geopolitical – advantage.

As with water, carbon and all other compounds on our planet, a truly circular economy is only possible by integrating the oceans and underwater life into the cycle to ensure its complete regeneration.

Seaweed for cleaning the oceans and revitalizing their biodiversity

Fortunately, initiatives seem to be emerging around the world. In the West, algae have been widely used for cleaning waste-water on land for a long time.[175] Filamentous algae (or fibrous algae and slimy alga) eliminate residual nutrients from the effluents of sewage treatment plants[176] and are then used as a biostimulant for plants.

In Australia, the government is developing seaweed barrier solutions to filter pollutants around the Great Barrier Reef. The challenge is to protect this ecological niche from the land-based pollution that is destroying this natural sea a little more every day.[177] The process is also well-known there: the high presence of nitrogen leads to a higher growth of micro- and macroalgae competing for light and mineral nutrition with the seaweed that ensure the symbiosis of the corals. Nitrogen also favours the development of starfish, large coral predators that are contributing to the disappearance of the Great Barrier Reef by devouring it. To counteract these phenomena, the cultivated seaweed captures nitrogen and prevents both starfish, microalgae and other macroalgae attached to the corals from proliferating. The high uptake of suspended nutrients by the seaweed will also help to reduce the turbidity of the water and thus increase the level of light reaching the corals to encourage their growth.

But the growth of corals and coralline algae that strengthen the reefs is also severely limited by ocean acidification. This is mainly due to excessive concentrations of anthropogenic carbon in the seas. While it is unthinkable that enough seaweed could be cultivated to deacidify the entire ocean, its presence will help to absorb a large amount of carbon dioxide from the surrounding area. This capture would form a kind of halo around the seaweed, that would protect the adjoining corals.

To make sure we mention all the benefits of these systems involving seaweed, it should be emphasized that besides their own potential, these algae also harbour many allies that make it possible to clean up the oceans and repair an already damaged biodiversity. Seaweed beds are marine habitats containing a large number of living organisms. The kelp's simple holdfast that tethers the seaweed to the ground contains hundreds of different organisms. Worms, crabs, molluscs and shrimps all live here together, forming an essential ecosystem.[178]

There are relatively few strictly vegetarian seaweed-eating fish, but many of them suck on the fronds to collect the organisms that develop there. Decomposed seaweed in the ocean also feeds the first links in the food chain, namely certain shellfish and a whole microbial chain that recycles nutrients to feed phytoplankton, which in turn helps to feed the fish.

The importance of phytoplankton (plant plankton) should be highlighted here. It represents 50% of the organic matter produced on earth and has been steadily decreasing over the last twenty years, partly due to human activities.[179] Phytoplankton is the largest source of oxygen for our planet and ensures the oxygenation of aquatic masses which without its action would become sterile. As it dies, it sinks into the sediments, where it also feeds the first level of the food chain, allowing it to survive and to purify the oceans itself.

At the same time, a single mussel is capable of filtering 25 litres of water per day and retaining pesticides, bacteria or drug residues. An oyster can filter up to 190 litres per day. This direct or indirect filtering capacity provided by seaweed is a considerable ecosystem service. It should be noted that through sedimentation, dead phytoplankton also contribute to the sequestration of carbon in the deep sea for millennia.

Seaweed cultivation can thus improve the function and structure of marine communities by serving as food, refuges and nurseries at every level of the marine life chain. Renewing this biodiversity at plant and animal levels will in turn provide important ecosystem services.[180]

Finally, on the long list of seaweed's services rendered, we must also remember its potential for reducing plastic pollution, absorbing carbon, reducing the need for chemical pesticides, or reducing ruminants' methane production. There is a huge range of possibilities. We urgently need to review our understanding of the role of seaweed and, above all, to consider ways of making the most of the ecosystem services it provides.

These efforts need to be supported and accompanied by scientific research, which must remain a safeguard and an absolute prerequisite for these developments.

Invasive seaweed

So as to avoid any misunderstanding, we should make it clear that not all existing phenomena of invasive seaweed causing imbalances are solely due to pollutants or eutrophication. Many ecosystem disturbances are related to opportunistic types of seaweed that grow in a new environment from which they were previously absent. Criss-crossed by ships filled with ballast water, exploited by aquaculture and connected by canals, the seas and oceans today form a vast network that inevitably sees the displacement of potentially invasive species capable of proliferation. Going against these processes that have always existed is hard. This is even more true given that the balance is now being modified by global warming.

These disruptions are also sometimes caused by people trying to introduce seaweed into new environments without taking the necessary precautions. There are famous precedents for this. One example is *Caulerpa taxifolia* in the Mediterranean. Nicknamed 'the green plague' or 'the killer algae', it appeared around Monaco in 1984. Suspicions were raised that the Oceanographic Museum of Monaco, like many tropical marine aquariums around the world, housed the Asian seaweed in its aquariums and may have accidentally released it into the wastewater. The Museum has always denied this.

In any case, within six years, carried by fishermen's nets or yachtsmen's anchors, *Caulerpa* had invaded 15,000 hectares of France, Spain, Italy, Croatia and Tunisia, replacing many local species including the famous *Posidonia oceanica* seagrass. *Caulerpa's* vegetative reproduction by cuttings made its development very rapid and difficult to control. It seems that the local ecosystems have now ended up self-regulating its presence over the years and it cannot be found now in significant quantities. This species of *Caulerpa* has been superseded by the abundant Red Sea *Caulerpa racemosa*, which has steadily colonized the Mediterranean via the Suez Canal. No one talks about this species, but it is far more worrying for Mediterranean biodiversity than the one the media was obsessed with in the 1980s.

This is not a unique experience. Even *wakame* has demonstrated an invasive, opportunistic character. In New Zealand, this phenomenon has forced the authorities to take drastic measures against it. In France, its introduction in the 1970s created major controversies which led to a freeze on new licences, after it was found to be spreading opportunistically. The subject of licences for *wakame* cultivation remains at the heart of a very lively debate in Brittany today.

The risks associated with the cultivation of a non-endemic species of seaweed are a key issue because controlling the propagation of an organism in the ocean is much more complex than on land and its consequences are more dramatic.

Any domestication of a seaweed in a new area must include extremely rigorous standards of quarantine beforehand. Besides the visible organism that is introduced, many invisible microbes can be carried with it and can wreak havoc on local life. This phenomenon has been evident to everyone for a long time, with the spread and transfer of diseases on land.

Industry experts must work to rapidly define these new environmental standards and could draw heavily on the Asian experience. Hopefully these principles will soon be taken up by international institutions such as the UN. Beyond the equality of regulatory requirements around the world, it is clear to all that the spread of seaweed knows no borders and cannot be regulated by purely national rules. Acting in coalition on these biosafety standards will minimize risks and facilitate the issuing of permits by the authorities, while sharing knowledge. These standards will also help to reassure and protect investors. As the experience of PepsiCo in India has shown, the lack of knowledge or non-definition of environmental risks opens the door to controversies and accusations that then become difficult to adjudicate.

Seaweed makes it rain

In order to take a broader view of the environmental impacts of our miraculous seaweed, we can, for a change, consider what it also produces in the sky. Clouds, like seaweed, have no regard for our territorial boundaries. And as surprising as it may seem, their fates are linked. We are in the early stages of researching these events. Nevertheless, they all demonstrate that iodine stored in large brown seaweed influences the coastal climate and contributes to cloud formation.[181] At low tide, kelp is dehydrated and exposed to ozone, and releases iodine gas into the air. This iodine then reacts with ozone under the effect of light to create fine particles that act as a condensation point for water vapour. This process forms droplets and thus clouds, and in the end, rain. Other gases – this time sulphurous – emitted by seaweed but also in large quantities by phytoplankton, contribute to the condensation of clouds through chemical processes linked to the formation of fine particles.

These are real virtuous cycles for iodine and sulphur: when the weather is good, photosynthesis works at full capacity and produces a lot of seaweed that forms large quantities of particles, which then generate clouds, rain, and then a drop in emissions from the seaweed, which clears the sky; then the weather is good again and the cycle continues...

Hopefully, our mastery of seaweed processes will one day prove more effective than animistic dances and chants for bringing rain in desert regions.

Apart from the benefits for arid regions, it is also worth remembering that water vapour now accounts for about 75% of greenhouse gases, way ahead of carbon and methane.[182]

Without water vapour, the temperature would be -20°C on earth. The problem we have today is that the water cycle is extensively damaged by disturbances from human activities. The depletion of soils unable to hold water, the massive extraction of underground water reserves and changes in our oceans have destroyed the natural ability of our atmosphere to create rain and to return the water vapour underground, so it stays in the atmosphere. The evolution of the concentration of water in the atmosphere in the long term is hard to predict, but everything suggests that a 'pressure cooker' effect is taking place, which is greatly accelerating climate change. It should be noted in passing that the climate benefits of nuclear energy – a major consumer of water for cooling the power plants – are becoming much more relative.

Some people say that carbon is just the tree that hides the forest of global warming. With this in mind, the ability of seaweed to generate rain and partially 'repair' this damaged water cycle could prove decisive for our future, both in terms of the state of our water reserves and the rise in the earth's temperature.

But influencing precipitation is not the only benefit of these iodine gas emissions. As Philippe Potin, director of research on algae at the CNRS, often says, this mechanism also cuts ozone concentrations by half, thus contributing to air quality on the coasts where seaweed production is abundant. This system is so efficient that it could be used to reduce the levels of ground-level ozone which is a harmful pollutant for humans, animals and plants. Again, immense precautions must be taken to ensure the safety of these approaches as these gases, if not neutralized in the lower atmosphere, would also contribute to altering our protective ozone layer in the upper atmosphere. It is important to consciously assess the adverse effects of a solution, even when it seems miraculous.

New seascapes

In all circumstances, seaweed cultivation will never be neutral and will cause complex environmental impacts. Some naturalists are concerned above all with preserving the existing biodiversity, and they are right to alert us to the risks of domestication to the detriment of wild species. Similarly, there is no doubt that fields of wheat and maize have replaced the many and varied wild grasses in the pastures. But the ocean of the nineteenth-century will never return and it is already impossible to maintain it in its present state. Human population has increased dramatically over the past few centuries, and no one can bring themselves to wish for a sudden drop in population, whether through war, hunger or pandemics. Our civilization has domesticated the earth's environment in order to survive, largely modifying it and creating the Anthropocene epoch. It is misguided to think that we can reverse this era and return to the wild ocean that has caused sailors to dream so much. The earth will continue to spew waste, people will continue to fish and the climate will continue to get warmer.

All of this is already out of our hands! Our marine ecosystem as we know it today has only a few years to live, and it will be profoundly modified in the coming decades. Managing the life cycle of the elements on earth without including two thirds of our blue planet is doomed to failure.

Civilization is characterized by the optimization of a number of parameters which inevitably include environmental elements. Our civilizations – sometimes out of idealism, often out of naivety – have so far decided not to really interfere in these great maritime spaces that seem so foreign to us. The idea of sustaining life in the oceans, which implies organized human intervention in these ecosystems, still provokes indignant reactions from many pessimists who believe that we are condemned to repeating the same mistakes. But we can still hope that we can learn from the past...

While preserving wilderness areas, we will also need to plant seaweed to clean the oceans and continue the cycle of carbon, phosphate, nitrogen, sulphur, water and other elements required for life on earth. There are bound to be risks, especially in an environment as complex as the ocean, but nothing seems riskier than allowing the planet to continue to collapse.

Algae are the base of the pyramid. We cannot ignore them any longer. The aim is to conserve their diversity and to restore the balance that is deteriorating. The species will no longer be exactly the same, the landscapes will change, and so will the balances. This underwater life will be different, but it will have at least one merit: it will still exist...

Enteromorpha intestinalis ('sea lettuce', 'green bait weed', 'gutweed' or 'grass kelp') – a green seaweed assimilated to a type of *Ulva*. Its 10 to 15 cm long thallus is swollen and covered with gas bubbles. It is almost fluorescent green in colour and stands upright in the water in the shape of an intestine, hence its name. It is found in abundance throughout the world, except in very cold waters. Rich in phosphorus and nitrogen, it was often used for spreading. It contains many vitamins and proteins. It is cultivated and consumed in Asia where it is simply called *aonori* (green seaweed). It often covers areas several kilometres long on beaches. Numerous marine invertebrates such as molluscs and crabs live in it.

9

CULTIVATING THE OCEANS IN THE WORLD OF 2050?

We were invited on a radio programme recently and the journalist asked the speakers to sum up seaweed in one word. Not two, not three: just one! The oceanographer sitting next to me thought for a moment, and then chose a very simple word: *hope*.

To what extent can seaweed represent a new hope for the world of tomorrow? Each of us holds part of the answer. It is our joint responsibility to turn this hope into a reality. To dream of a world where the economy's only aim is to repair ecosystems and improve social justice is perhaps utopian. However, in the past, the 'utopians' have made many advances that seemed far-fetched to their contemporaries.

Two hundred years ago, when the earth had its first billion inhabitants, Thomas Malthus and his loyal followers predicted that it would be impossible to find enough resources on the planet to feed such a large population. In 1950, two thirds of people were dying of hunger, while the world's population stood at two billion.

Over the past seventy years, the 'utopians' have developed food systems where only one in nine people are hungry, even though our global population has grown to eight billion. At the same time, the 'utopians' have fought to achieve higher literacy rates, gender equality, greater tolerance of minorities, the establishment of democratic regimes in half of the world's countries and a historic reduction in the number of armed conflicts.

We owe a lot to those who have chased this utopia.

In this chapter, let's become ocean utopians: what would the world be like in 2050 if we fully integrated seaweed, and the ecosystem it supports, into the way it functions?

It's New Year's Eve 2050, and seaweed features on most dinner menus. We are now aware of the benefits of seaweed, both for the environment and for our health. Recent research into our gut microbiota supports these findings. Our eating habits, respect for the environment and knowledge of our bodies have changed enormously. Seaweed is also trendy among young people, who are increasingly turning more towards a plant-based diet. If in the 1950s, rock music, westerns and Hollywood gave us American burgers and sodas, since the 2020s, Japanese manga, Chinese soft power and K-pop have sprinkled seaweed onto our plates. Lactofermentation makes the taste of seaweed more accessible to the Western public. This fermentation also makes it possible to avoid systematically drying it, which is a very energy-consuming process.

To meet the growing demand, seaweed cultivation has expanded massively throughout the world and far beyond the Asian continent. The European and American governments were the first to invest seriously in the development of a sector using local species. Specific efforts have also been made to facilitate access to cultivation areas and to accelerate the distribution of offshore concessions. At the same time, the rapid development of offshore wind farms has increased opportunities for cultivation in areas previously untouched by production.

Investments at the global level have accelerated since the UNFCCC (UN Framework Convention on Climate) introduced carbon offset mechanisms linked to seaweed, after having quantified the impact of this 'blue carbon' on climate change. This recognition and promotion of the role of seaweed for the environment has naturally facilitated access to funding. Many young entrepreneurs have decided to invest in the seaweed industry. Increasing numbers of them are now making a good living from their new profession, despite the high initial outlays and the time it takes to turn a profit.

International standards have been rapidly adapted through traceability systems and a standardization effort on the part of international institutions. All this progress has received firm support since the creation in 2028 of UN-Oceans, the first UN agency for the collective management of the high seas. This decision was initially seeded during the UN-Oceans Conference 2025 in Paris.

In India and Africa, seaweed cultivation allows for greater food sovereignty, which reduces the nutritional deficiencies of the population and dependence on international aid. Algaculture is developed, has its own curriculum in universities and integrates emerging aquaculture activities. Large areas of regenerative permaculture at sea are emerging on the coasts. Ambitious training programmes for coastal communities – financed by major international donors and non-governmental organizations – have anticipated the disappearance of non-artisanal fisheries and reoriented people towards aquaculture using seaweed.

In recent years, huge floating seaweed fields have been observed offshore by solar-powered drones flying over the upwelling areas close to Senegal and Namibia. These new landscapes have become a symbol of Africa's burgeoning capacity for innovation.

In the east of the continent, coastal countries have developed new species while integrating other animal or shellfish crops. A large global movement called 'She-weed' has also emerged on the African continent, using this new industry to fight against the old patriarchal models of aquaculture and fishing. Active on the latest forms of social media, this virtual community brings together women from all over the world and launches innovative actions.

The Indian government has made it compulsory for aquaculture production, particularly shrimp production, to incorporate seaweed in order to limit waste runoffs and reduce negative effects on the environment.

Ecological awareness has gradually spread to all countries around the world.

Further north, global warming and ice melt have opened up large areas of production in northern Siberia, Alaska, Greenland and Canada. In two decades, these countries have become major players in the seaweed world. The rapid development of sectors in these areas has attracted new populations and marked the beginning of what some call the 'Cold Blue Rush', referring to the California Gold Rush of the nineteenth century.

Thanks to the contribution of seaweed and integrated aquaculture products, food insecurity has been drastically reduced and has almost disappeared by the early 2040s. The remaining pockets of famine and nutritional deficiencies are mainly due to local conflicts or very specific political or logistical difficulties. In addition, the increasing use of seaweed protein extracts in animal feed, instead of GM soybean meal from Brazil, has been a major factor in halting the massive deforestation of the Amazon. More generally, seaweed cultivation, together with the strong development of vegetarian diets, has freed up land for less intensive livestock and other crops.

Underwater farming has seen rapid advances in machine learning combined with the Internet of Things and the widespread use of sensors at sea that can guide growers based on the microbial properties of the water. This has resulted in a real capacity to prevent external contaminations and aggressions in these seaweed fields, which are now remotely controlled via satellite monitoring. The creation of coalitions of academic and industrial actors working in collaboration has greatly accelerated the dissemination of this new knowledge. Special 'seaweed farming drones' that are designed to respect ecosystems have been flying over the beaches since the 2030s.

The profit margins on the sale of edible seaweed also improve the sophistication of biorefineries for transforming certain sea-weeds into several by-products. Marine plants are used as natural fibres in textiles and are gradually replacing cotton, which requires a large quantity of water and pesticides. Seaweed extracts are also helping to replace plastics, which will almost certainly be banned by 2043. The introduction of a 'plastic neutrality' obligation in the 2030s forced major plastic-using industries to offset their plastic use by financing innovative projects. These projects have enabled the development of alternatives to plastic and, more generally, more sustainable value chains based on marine biotechnology. Research into algal microbiology has become a cutting-edge sector and pharmaceutical companies are increasingly incorporating seaweed compounds into the manufacture of medicines. This new capacity is accompanied by advances in nanotechnology and immunotherapy. Thus, since the 2040s, a real drop in cancer rates has been observed as preventative and curative treatments based on seaweed increase in efficacy.

Other molecules from Kappaphycus *have been significant in curbing the Ebola-37 pandemic. Less than twenty years after the Covid-19 crisis, the mutation of this haemorrhagic fever from Africa once again shook the world in 2037. The vaccine finally established against Ebola-37 uses RNA genes from the alga* Laminaria digitata, *while nasal sprays have been developed based on alginate, to replace face masks.*

By effectively stimulating animals' immune systems and growth, seaweed has enabled a reduction in the use of veterinary antibiotics by 90% between 2020 and 2040. This decrease probably explains why, against all expectations, antibiotic resistance in humans declined over the same period. Thus, the antibiotic molecules of 2020 are still efficient today.

In addition, by 2042, the combined action of fermented and non-fermented seaweed has almost completely eliminated methane emissions from ruminants worldwide. This is a major development, accompanied by incentives from both governments and the beef industry worldwide.

The huge invisible fields of seaweed all along our coastlines are now able to absorb and store large amounts of carbon. These two factors contribute significantly to slowing down global warming and to achieving, despite the predictions, the commitments made at COP 32, where the original Paris agreements were revised.

At that time, the first open ocean crops were launched, piloted by aquatic drones capable of submerging the seaweed nets at night to fetch nutrients from the sediment and then raising them during the day in order to capture sunlight. Other systems send in underwater drones that use tidal energy to bring nutrients from the shallows up to the level of the seaweed.

Wild seaweed, which is well preserved in large, dedicated areas, is now monitored and protected by satellite in order to revitalize the plant biodiversity of the oceans.

This rapid multiplication of production caused initial inflation in the sector. The production of sea vegetables has skyrocketed, flooding the markets and outstripping demand. In order to use this situation to combat global warming, UN-Oceans decided in 2043 to turn some seaweed farms into gigantic carbon pumps designed to sequester carbon at the bottom of the oceans, hoping therefore to actively cool the atmosphere. In order to prevent the system from drifting, aquatic drones have been put in place to monitor the deep seabed, which hosts seaweed on bacterial mats designed to accelerate their sedimentation in the abyssal plains. In addition, as lipid extraction capabilities greatly improved in the 2030s, some microalgae have also been used to create bioethanol and replace the remaining fossil fuels. Thanks to tidal energy, submerged machines were also created in the same period, replicating the movements of seaweed in the water to produce energy.

All of these advances, combined with other carbon austerity measures, have helped to avert the climate catastrophe that was heralded since the beginning of the century. In recent years, there has even been a slight decrease in the level of greenhouse gases in the atmosphere, which would indicate a possible cooling in the coming years.

On land, seaweed is still used as a biostimulant for plants, especially in areas of high pollution, where it is actively involved in coastal clean-up and the recycling of land-based waste. Huge seaweed fields are now being installed at the mouths of the Amazon and major rivers to reuse the pollutants that used to accumulate there. Natural sea-purification plants have been installed in the world's most polluted bays. These installations have naturally put an end to the destructive seaweed tide phenomena.

The phosphate crisis of the mid-2030s, which began as a result of a depletion of mineral resources and should have led to major global famines, was largely resolved by these pollutant-recycling systems. Growing these crops, which are capable of recovering phosphate compounds from fertilizers at sea and reusing them, has enabled the creation of a regenerative and circular agriculture between land and sea.

Most recently, the increased sophistication of techniques using iodine from seaweed, combined with the launch of tropospheric drones and nanotechnology to seed clouds, has increased our understanding of the entire water cycle on earth. It is therefore now possible to transport rainwater to desert regions. The harmonious distribution of water resources on the planet has greatly reduced drought phenomena. The emergence of new arable land, coupled with increasing ocean resources, now makes it possible to sustainably feed a growing and ageing population that is increasingly less prone to disease.

So, in less than thirty years, humans have succeeded in mastering the complete cycle of the three major essential compounds: water, carbon and phosphate.

For the first time in our history we have managed to feed our entire population properly.

And this success has not been at the expense of the planet; quite the contrary. Next year, the first installations of offshore cities will be launched. Huge barges several square kilometres in size will be installed several kilometres from our coasts, in regions where the population density is becoming too high. The inhabitants of these new marine cities will be able to feed themselves locally with seaweed, fish and shellfish cultivated in an integrated and regenerative way. Tidal, wind and solar energy will provide the electricity needed for the agglomeration. Fresh water will be made available through spin-drying the large kelp, whose structures naturally desalinate the sea water. The first offers to live on these futuristic platforms were a great success and many new projects are under consideration.

The long-duration space travel that has been democratized over the last ten years also widely uses dried seaweed to feed astronauts and provide them with a number of necessary raw materials on their long journeys. NASA's old concept from the 1970s – that carbon released by astronauts' exhalations could be rapidly converted into edible biomass – is now finally operational, and the use of molecules extracted from Macrocystis *is now becoming commonplace. If this process is firmed up and becomes widespread, the cooling of the atmosphere could accelerate. Studies are underway to replicate the marine ecosystem on other planets using microalgae.*

And the best is yet to come…

Obviously, this whole projection is pure fiction. The future will undoubtedly be very different and more surprising than anyone could imagine.

Who, in the early 1990s, listening to their cassette player or inserting a floppy disk into their computer, could have predicted the digitalization of our lives and our virtual memories, intertwined at our fingertips on phones more powerful than any computer at the time?

The only certainty is the acceleration of change.

This projection is interesting because it is feasible. Idealistic for some, or nightmarish for others, it is within the limits of what is possible with our current scientific advances.

The most uncertain part of this prediction is our common desire to implement this integration of the oceans to help solve the major challenges of our generation.

We are all connected to the same system and we are all involved in this decision. What role will we give to seaweed?

No doubt there will still be many sceptics who will see this seaweed revolution as nothing more than some bizarre flight of fancy. But although it may be impossible to predict the future of our humanity, it is very easy to recount its past. Life originated in the ocean billions of years ago and only very recently began to evolve outside the water. Our connection to the sea is unique and timeless. As a condition of our life on earth, our relationship with this immense expanse, which we believe to be inexhaustible, could well decide the future of our civilizations.

An ocean view fills us with serenity. Seeing the sea, swimming in it, immersing oneself in seawater and sunshine is a joy for every human being. The human body is essentially made up of water. The ocean is not part of our environment, we are part of the oceans' environment!

As a father of three, I cannot bring myself to watch those already being called 'Generation Covid' experience the predicted extinction of our species. This fight for seaweed is above all an intergenerational commitment to hope and optimism. We could choose to forget where we came from and merely survive while we wait for global warming and the other cataclysms predicted by the media pundits who ride the wave of our fears.

'We need to move beyond guilt or blame, and get on with the practical tasks at hand,' as David Attenborough says.[183]

We are feeding our children with fear and drama when we should be fighting to provide them with solutions and a source of optimism. There is hope. And judging by the visionary enthusiasm of these seaweed pioneers joining us in increasing numbers every day, it is possible to believe.

We believe that all this ongoing research into seaweed and the projects that have been launched, both locally and globally, hold great promise for the future.

We believe in a future where world hunger, pollution, certain diseases, climate change, loss of biodiversity, social inequality and impoverishment will be reduced.

We believe that by reconnecting with living things and collaborating with the marine prodigies with which we share our origins, we can bring about an ecological, geopolitical, medical, energetic, social and humanist revolution.

The seaweed revolution.

Acknowledgements

A huge thank you to my 'brother in algae', the marine biologist Philippe Potin, CNRS research director at the Roscoff Marine Station, who has taught me so much about these organisms over the past few years and has been kind enough to read over these pages. This book would never have existed without him, without his unique generosity and his encyclopaedic knowledge of seaweed, a knowledge that is matched only by his enthusiasm for sharing it.

Thanks also to my companions who had this visionary belief since the very first moment, Ruth, Tim, David, Beth and Olivia from the Lloyd's Register Foundation 'family' as well as the entire UN Global Compact Ocean Team and especially to Erik and Marta for their support. Thanks to Nichola, Azzedine, Kevin and Sofya for supporting the early days of our new seaweed coalition.

Special thanks to Jeanne Pham Tran, for believing that I could write a book about this topic, and for supporting and editing my initial forays into writing about seaweed.

More than anything, I'd also like to thank all those who agreed to contribute to this book, enriching it with their knowledge and their experience, in particular: Alan Critchley, Flower Msuya, Stefan Kraan, Helena Abreu, Alejandro Bushmann, Yoichi Sato, Gwan Hoon Kim, Junning Cai, Jorunn Skjermo, John Bolton, Jo Kelly, Charles Yarish, Al-Jeria Abdul, Adrien Vincent, Liu Tao, Ian Neish, Olavur

Gregersen, Liz Cottier-Cook, Carlos Duarte, Mounir Boulkout, Thierry Chopin, Briana Warner, Sander Van Der Burg, Bren Smith, Anton Voskoboynikov, Housam Hamza, Rhianna Rees, Nancy Iraba, Pierre Paslier, Klaartje Schade, Paul Dobbins, Michael Roleda, Daniel Hooft and all the other pioneers, the architects of the changes to come...

We will owe this revolution to their passion and hard work!

Notes

1 The term 'seaweed' is used to refer to 'macroalgae', which is the focus of this book. The term 'algae' refers to both micro- and macroalgae.

2 Microalgae are generally photosynthetic micro-organisms, eukaryotic or prokaryotic (with or without a cell nucleus) and play a fundamental role in the carbon cycle and, more generally, in the biogeochemical cycles of lakes and the ocean. They are cultivated on land, and although their production is highly valued and has great potential, it remains insignificant to date (250 tonnes worldwide).

3 Duarte, C.M., Bruhn, A., Krause-Jensen, D., 'A seaweed aquaculture imperative to meet global sustainability targets', *Nature Sustainability* (2021).

4 Food and Agriculture Organization of the United Nations, 'Seaweeds and Microalgae: an overview for unlocking their potential in global aquaculture development', ISSN 2070-6065 (2021).

5 www.fao.org/3/cb5670en/cb5670en.pdf.

6 www.atlasobscura.com/foods/cochayuyo-seaweed-chile.

7 Dillehay, T.D., 'New Archaeological Evidence for an Early Human Presence at Monte Verde, Chile', *PLOS One* (Nov. 2018).

8 Lindo, J. et al., 'The genetic prehistory of the Andean highlands 7000 years BP though European contact', *Sciences Advances* (Nov. 2018).

9   Erlandson, J.M., et al., 'The Kelp Highway Hypothesis: marine ecology, the coastal migration theory, and the peopling of the Americas', *Journal of Island and Coastal Archaeology*, Vol. 2, Issue 2 (2007).

10 Crawford, M.A., Bloom, M., Broadhurst, C.L., Schmidt, W.F., Cunnane, S.C., Galli, C., Gehbremeskel, K., Linseisen, F., Lloyd-Smith, J., Parkington, J., 'Evidence for the unique function of docosahexaenoic acid during the evolution of the modern hominid brain', *Lipids* (1999).

11 University of Southern Denmark, 'Did seaweed make us who we are today?', *Science Daily* (February 2017).

12 United Nations, 'Report on Standards Projection World Population prospects' (2019) population.un.org/wpp/Download/Standard/Population/.

13 Food and Agriculture Organization of the United Nations, *How to Feed the World in 2050* (2009).

14 'Food in the Anthropocene: the EAT–Lancet Commission on healthy diets from sustainable food systems', *The Lancet Commissions*, Vol. 393, Issue 10170, (February 2, 2019), p. 447–492, available at www.thelancet.com/journals/lancet/article/PIIS0140-6736(18)31788-4/fulltext.

15 Bajželj, B., Richards, K., Allwood, J. et al., 'Importance of food demand management for climate mitigation', *Nature Climate Change 4* (2014).

16 Schubel, J.R., Thompson, K., 'Farming the Sea: The Only Way to Meet Humanity's Future Food Needs', *GeoHealth*, Vol. 3, Issue 9 (August 2019).

17 Hélène Marfaing, Report by CEVA (Centre d'étude et de valorisation des algues [Algae Technology and Innovation Center]). *Algues. Production mondiale et usage* (2021).

18 Zava, T.T. and Zava, D., 'Assessment of Japanese iodine

intake based on seaweed consumption in Japan: A litera-ture-based analysis', *Thyroid Research*, Vol. 4 (October 2011).

19   Heilmayr, R., Rausch, L., Munger, J. and Gibbs, H., 'Brazil's Amazon Soy Moratorium reduced deforestation', *Nature Food* (December 2020).

20   CEVA documentation, 'Nutritional composition of sea-weed' (2021). Available at www.ceva-algues.com/en/document/nutritional-data-sheets-on-algae/.

21   The World's 50 Best Restaurants (2019). Available at www.theworlds50best.com/the-list/1-10/Mirazur.html.

22   Examples of cookery websites:
      irishseaweedkitchen.ie/
      maraseaweed.com/blogs/recipes
      atlanticseafarms.com/blogs/news
      www.greenwave.org/recipes.

23   Examples of seaweed cookery books:
      The Limu Eater: A cookbook of Hawaiian seaweed by Heather J. Fortner (Originally published in 1978 and a second printing was recently updated and published by Dr. Celia Smith and Pelka Andrade).
      Madlener, J.C. 1977. *The Sea Vegetable Book*. New York: Clarkson N. Potter, Inc.
      Rhatigan, P. 2009. *Irish Seaweed Kitchen*. Booklink. www.prannie.com
      Rutt, J. and S. Mattielli, 1990. *Lee Wade's Korean* Hollym International Corp., Elizabeth, NJ. Cookery
      Tsuda, N. 1998. *Sushi Made Easy*. Weatherhill, Inc., New York.
      The Seaweed Cookbook: A guide to edible seaweeds and how to cook with them: amzn.eu/d/7DivInz

The Seaweed Cookbook: Discover the health benefits and uses of seaweed, with 50 delicious recipes: townsend-sofmanningtree.co.uk/products/he-seaweed-cookbook-discover-the-health-benefits-and-uses-of-seaweed-with-50-delicious-recipes.

The Seaweed Cookbook – The Cornish Seaweed Company: www.thestivesco.co.uk/products/cornish-seaweed-cookbook.

24 The association of algae with a colour is a botanical concept which in some circumstances differs from the actual colour. For example, some red algae will often be green during their lifetime, while others will be yellow, purple, pink, etc.

25 British Chambers of Commerce Korea, *Market Research Report Sea Fish Industry Authority* (November 2018).

26 Food and Agriculture Organization of the United Nations, *Seaweeds and Microalgae: an overview for unlocking their potential in global aquaculture development*, ISSN 2070-6065 (2021).

27 Berche, P., 'L'histoire du scorbut', *La Revue de biologie médicale*, No. 347 (March 2019).

28 Cook, J., *The Explorations of Captain James Cook in the Pacific, as Told by Selections of his Own Journals 1768-1779*, Grenfell Price (1971).

29 Nielsen, C.W., Rustad, T., Holdt, S.L., 'Vitamin C from Seaweed: A Review Assessing Seaweed as Contributor to Daily Intake', *Foods* (2021).

30 Food and Agriculture Organization of the United Nations, *Kimchi, seaweed, and seasoned carrot in the Soviet culinary culture: the spread of Korean food in the Soviet Union and Korean diaspora* (2016).

31 Hehemann, J.H., Correc, G., Barbeyron, T., et al., 'Transfer of carbohydrate-active enzymes from marine bacteria to Japanese gut microbiota', *Nature* (2010).

32 Fleurence, F., *Les Algues alimentaires : bilan et perspectives*, Lavoisier, coll. Sciences et techniques agroalimentaires (2018).

33 Safe Seaweed Coalition, available at www.safeseaweedcoalition.org.

34 Williams, A.G., Withers, S., Sutherland, A.D., 'The potential of bacteria isolated from ruminal contents of seaweed-eating North Ronaldsay sheep to hydrolyse seaweed components and produce methane by anaerobic digestion in vitro', *Microb Biotechnol.* (January 2013).

35 Food and Agriculture Organization of the United Nations, 'Report: The State of World Fisheries and Aquaculture' (2020).

36 Fry, J.P. et al., 'Feed conversion efficiency in aquaculture: do we measure it correctly?', *Environ. Res. Lett.* 13 (2018).

37 Ahmed, N., Thompson, S., Glaser, M., 'Integrated mangrove-shrimp cultivation: Potential for blue carbon sequestration', *Ambio* (May 2018).

38 Hui, Y., Tamez-Hidalgo, P., Cieplak, T., et al., 'Supplementation of a lacto-fermented rapeseed-seaweed blend promotes gut microbial and gut immune-modulation in weaner piglets', *Journal of Animal Science and Biotechnology*, 12 (2021).

39 O'Neill J., 'Review on Antimicrobial Resistance, Tackling drug resistant infections globally: final report and recommendations', *AMR* (2016).

40 Calder, P.C., 'n-3 polyunsaturated fatty acids, inflammation, and inflammatory diseases', *Am. J. Clin. Nutr.* (June 2006).

41 Simopoulos, A.P., 'Evolutionary aspects of diet, the omega-6/ omega-3 ratio and genetic variation: nutritional implications for chronic diseases', *Biomed. Pharmacother.* (November 2006).

42 Su, K.P., Huang, S.Y., Chiu, C.C., Shen, W.W., 'Omega-3 fatty acids in major depressive disorder. A preliminary double-blind, placebo-controlled trial', *Eur. Neuropsychopharmacol.* (August 2003).

43 Pereira, H., Barreira, L., Figueiredo, F., Custódio, L., Vizetto Duarte, C., Polo, C., Rešek, E., Engelen, A., Varela, J., 'Polyunsaturated Fatty acids of marine macroalgae: potential for nutritional and pharmaceutical applications', *Marine Drugs* (September 2012).

44 US Environment Protection Agency, 'Global Greenhouse Gas Emissions Data 2021', available at www.epa.gov/ ghgemissions/global-greenhouse-gas-emissions-data.

45 Food and Agriculture Organization of the United Nations (FAO), 'Global Livestock Environmental Assessment Model (GLEAM)' (2021). Available at www.fao.org/gleam/en/.

46 Ritchie, H., 'Our World in Data: Cars, planes, trains: where do CO_2 emissions from transport come from?' (June 2020), available at ourworldindata.org/ co2-emissions-from-transport.

47 Roque, B.M., Venegas, M., Kinley, R.D., de Nys, R., Duarte, T.L., Yang, X., Kebreab, E., 'Red seaweed (*Asparagopsis taxiformis*) supplementation reduces enteric methane by over 80 percent in beef steers', *PLOS One* (March 2021).

48 United Nations Environment Programme (UNEP), 'Report UNEP/Global Methane Assessment: Benefits and Costs of Mitigating Methane Emissions' (May 2021).

49 Yakhin, O.I., Lubyanov, A.A., Yakhin, I.A., Brown, P.H., 'Biostimulants in Plant Science: A Global Perspective', *Frontiers in Plant Science*, vol. 7 (2017).

50 www.seaweed.ie/uses_general/fertilisers.php.

51 Di Mola, I., Conti, S., Cozzolino, E., Melchionna, G., Ottaiano, L., Testa, A., Sabatino, L., Rouphael, Y., Mori, M., 'Plant-Based Protein Hydrolysate Improves Salinity Tolerance in Hemp: Agronomical and Physiological Aspects', *Agronomy* (February 2021).

52 ec.europa.eu/environment/nature/natura2000/index_en.htm.

53 Dizerbo, A.-H. and Floc'h, J.-Y., 'L'algue géante et le problème de son introduction – *Macrocystis pyrifera*', *Bretagne vivante* (1973).

54 Groupe Azote, *Rapport Commifer. Calcul de la fertilisation azotée* (2013).

55 Krause-Jensen, D., Lavery, P., Serrano, O., Marbà, N., Masque, P. and Duarte, C.M., 'Sequestration of macroalgal carbon: the elephant in the Blue Carbon room', *Biology Letters*, Vol. 14 (June 2018).

56 United Nations Global Compact (UNGC), 'Seaweed as a Nature Based Solution, Vision Statement' (October 2021), available at seaweedclimatesolution.com/.

57 Available at www.oceans2050.com/seaweed.

58 'Energy Futures Initiative, 'Uncharted Waters: Report on Expanding the Options for Carbon Dioxide Removal in Coastal and Ocean Environments', *Our Energy Policy* (December 2020).

59 Ritchie, H. and Roser, M., 'Our World in Data – CO and Greenhouse Gas Emissions' (2020), available at ourworldindata.org/co2-and-other-greenhouse-gas-emissions.

60 www.ogsociety.org/journal/.featured-articles/352-queen-of-weeds.html.

61 Schoenherr, A.A., *Natural history of the islands of California*, University of California Press (1999).

62 Gundersen, H., Rinde, E., Bekkby, T., Hancke, K., Gitmark, J. and Christie, H., 'Variation in Population Structure and Standing Stocks of Kelp Along Multiple Environmental Gradients and Implications for Ecosystem Services', *Frontiers in Marine Science*, Vol. 8 (2021).

63 Miller, R.J., et al., *Community structure and productivity of subtidal turf and foliose algal assemblages* (2009).

64 Buck-Wiese, H., Andskog, M.A., Nguyen, N.P., et al., 'Fucoid brown algae inject fucoidan carbon into the ocean', *Proceedings of the National Academy of Sciences of the United States of America*, 120(1) (January 2023).

65 www.seafoodsource.com/news/environment-sustainability/urchin-farming-company-gets-world-first-blue-carbon-credit-for-kelp.

66 US Dept of Health and Human Services, 'Iodine Fact Sheet for Health Professionals' (2021), available at ods.od.nih.gov/factsheets/Iodine-HealthProfessional/.

67 According to ANSES, the iodine level in iodized salt is 1860 mg/100 g compared to 1.8 mg/100 g for non-iodized salt (2019). Available at www.anses.fr/en/content/iodine.

68 World Health Organization, *Rapport sur les troubles dus à une carence en iode dans la région africaine de l'OMS : Analyses de la situation et perspectives* (September 2008).

69 Choudhary, B., Chauhan, O.P., Mishra, A., 'Edible Seaweeds: A Potential Novel Source of Bioactive Metabolites and Nutraceuticals with Human Health Benefits', *Frontiers in Marine Science*, Vol. 8 (2021).

70 Pruteanu, L.L., et al., 'Transcriptomics predicts compound synergy in drug and natural product treated glioblastoma cells', *PLOS One* (2020).

71 Woyengo, T., Ramprasath, V. and Jones, P., 'Anticancer effects of phytosterols', *Eur. J. Clin. Nutr.* (April 2009).

72 Roswell Park, *Breast Cancer Rates Rising Among Japanese Women* (July 2017).

73 OECD, 'Report on Obesity Update' (2017), available at www.oecd.org/els/health-systems/Obesity-Update-2017.pdf.

74 Available at www.bluezones.com/.

75 World Health Organization, 'Life expectancy and healthy life expectancy data by country' (2020). Available at apps.who.int/gho/data/node.main.688.

76 Leibbrandt, A., Meier, C., König-Schuster, M., Weinmüllner, R., et al., 'Iota-Carrageenan Is a Potent Inhibitor of Influenza A Virus Infection', *PLOS One*, 5 (12), e14320 (December 2010).

77 Shefer, S., Robin, A., Chemodanov, A., et al., 'Fighting SARS-CoV-2 with green seaweed Ulva sp. extract: extraction protocol predetermines crude ulvan extract anti-SARS-CoV-2 inhibition properties in in vitro Vero-E6 cells assay', *Peer Journal* (November 2021).

78 Morokutti-Kurz, M., Fröba, M., Graf, P., Große, M., Grassauer, A., Auth, J., et al., 'Iotacarrageenan neutralizes SARS-CoV-2 and inhibits viral replication in vitro', *PLOS One* (November 2020).

79 Stathis, C., Victoria, N., Loomis, K., et al., 'Review of the use of nasal and oral antiseptics during a global pandemic', *Future Microbiol.* (January 2021).

80 Smith, H.F., Parker, W., Kotzé, S.H., Laurin, M., 'Morphological

evolution of the mammalian cecum and cecal appendix',
Comptes Rendus Palevol, Vol. 16, Issue 1 (2017).

81 Neushul, P., 'Seaweed for War: California's World War
I Kelp Industry', *Technology and Culture*, vol. 30, no. 3,
(1989).

82 Émile Zola, *How Jolly Life is: A Realistic Novel* (1886),
translated from the 44th French Edition. Available at
books.google.co.uk/books/about/How_Jolly_Life_
is.html?id=duIzAQAAMAAJ&redir_esc=y.

83 United Nations Environment Programme (UNEP), *Banning
single-use plastic: lessons and experiences from countries*
(2018).

84 www.notpla.com.

85 Notpla, 'The Shadow Price of Plastic', available at
www.notpla.com/2022/07/21/the-shadow-price-of-plastic/.

86 www.loliware.com/.

87 eranovabioplastics.com/.

88 Immirzi, B., Santagata, G., Vox, G. and Schettini, E.,
'Preparation, characterisation and field testing of a biode-
gradable sodium alginate-based spray mulch', *Biosystems
Engineering* (2009).

89 World Wildlife Fund, *Report on the impact of a Cotton T
Shirt* (2013), available at www.worldwildlife.org/stories/
the-impact-of-a-cotton-t-shirt.

90 Combe, M., 'La production des OGM repart à la hausse
dans le monde', *Natura-Sciences* (March 2018).

91 bioplasticsnews.com/2019/07/31/irish-lidle-to-sell-seaweed-
underwear/.

92 renovaremx.com/.
mexiconewsdaily.com/news/eco-shoes-are-made-with-
sargassum-seaweed/.

93 vyldness.de.
www.greenqueen.com.hk/vyld-sustainable-menstruation-products/.

94 www.facebook.com/mucoalgue44.

95 De Morais Garofalo, L., , 'Les algues, des plantes marines peu utilisées au statut d'une «algarchitecture» (Seaweed, from little used marine plants to the status of an «algarchitecture»)', Master thesis, ENSA Paris-Malaquais, 2022.

96 terebess.hu/english/haiku/buson.doc.

97 www.wisdomportal.com/RobertBly/Bly-Haikus.html.

98 Éluard, P., poem, 'Où la vie se contemple tout est submergé'.

99 Prévert, J., poem, 'Démons et Merveilles', English translation available at www.babelmatrix.org/works/fr/Pr%C3%A9vert%2C_Jacques-1900/Sables_mouvants/en/33847-Quicksand.

100 Vivien, R., poem, 'En débarquant à Mytilène' – translator's own version.

101 Vivien, R., poem, 'Je pleure sur toi...' – translator's own version.

102 Vivien, R., poem, 'À la bien-aimée' – translator's own version.

103 Longfellow, H.W., poem, 'Seaweed', available at www.poetryfoundation.org/poems/44645/seaweed.

104 Sinatra, F., song, 'The Sea Song'.

105 gtep.technion.ac.il/electricity-from-the-sea/.

106 QualiKet Research, 'Rapport mondial sur la taille, la part, la demande et les prévisions du marché de l'alginate d'ici 2027', DuPont de Nemours, Marine Biopolymers Ltd, KIMICA, Algaia, Dohler Group, Ingredients Solutions, Ceamsa, Danisco A/S, FMC Corp. (January 2021).

107 www.hortimare.com/.

108 Msuya, F. and Hurtado, A., 'The role of women in seaweed aquaculture in the Western Indian Ocean and South-East Asia', *European Journal of Phycology*, 52 (October 2017).

109 Neish, I., Sepulveda, M., Hurtado, A., Critchley, A., 'Reflections on the Commercial Development of Eucheumatoid Seaweed Farming' (2017).

110 Food and Agriculture Organization of the United Nations (FAO). *Seaweed FAO Fact Sheet/Global seaweeds and microalgae production, 1950-2019* (2021).

111 Available at www.populationdata.net/palmares/idh/.

112 Marine Colloids will be acquired by DuPont, which recently sold its activities to International Flavors & Fragrances, one of the world's leading flavour companies. The French operations will be acquired by the American company Cargill. CP Kelco, still based in Denmark, remains a market leader in texturizers.

113 Translator's own version.

114 Msuya, F., 'The Impact of Seaweed Farming on the Social and Economic Structure of Seaweed Farming Communities in Zanzibar, Tanzania' (January 2006).

115 Eklöf, J., Msuya, F., Lyimo, T. and Buriyo, A., 'Seaweed Farming in Chwaka Bay: A Sustainable Alternative in Aquaculture?' (January 2012).

116 Msuya, F., 'The Seaweed Cluster Initiative in Zanzibar, Tanzania' (September 2006).

117 mwanizanzibar.com/.

118 www.foodingredientfacts.org/irish-moss-the-history-of-carrageenans-roots/.

119 healthyseaweedcafe.co.tz/.

120 International Union for Conservation of Nature (IUCN), *IUCN case study examines the interaction between*

aquaculture and marine conservation in Zanzibar (May 2020).

121 OECD (2021), available at sdg-financing-lab.oecd. org/.

122 Buschmann, A.H., Camus, C., Infante, J., Neori, A., Israel, A., Hernández-González, M.C., Pereda, S.V., Gomez-Pinchetti, J.L., Golberg, A., Tadmor-Shalev, N., Critchley, A.T., 'Seaweed production: overview of the global state of exploitation, farming and emerging research activity', *European Journal of Phycology* (2017).

123 Neish I., 'Social and economic dimensions of carrageenan seaweed farming in Indonesia', *Social and Economic Dimensions of Carrageenan Seaweed Farming* (2013).

124 Hurtado, A., 'Social and economic dimensions of carrageenan seaweed farming in the Philippines', *Social and Economic Dimensions of Carrageenan Seaweed Farming* (2013).

125 Hussin, H. and Khoso, A., 'Migrant Workers in the Seaweed Sector in Sabah, Malaysia', *SAGE Open*, 11, Issue 3 (2021).

126 Seadling Ltd, Sabah, Malaysia.

127 Murphy, D., 'Filipinos Swap Guns for Rakes', *Christian Science Monitor* (March 2002).

128 Kenicer, G., Bridgewater, S., Milliken, W., 'The Ebb and Flow of Scottish Seaweed Use', *Botanical Journal of Scotland* (2000).

129 Pays de Brest Algae Cluster, *Rapport d'étude : Poids économique de la filière algues en pays de Brest* (June 2021).

130 Burel, T., Le Duff, M. and Ar Gall, E., 'Updated checklist of the seaweeds of the French coasts, Channel and Atlantic Ocean', *An Aod – Les Cahiers naturalistes de l'Observatoire marin*, Vincent Le Garrec, Jacques Grall éd. (2019).

131 Updated estimate 2021 (Philippe Potin, CNRS), based on the IDEALG Project study on 'La filière des algues en France: évolution et poids économique' (2017).

132 Seaweed For Europe, 'Investor Memo/The case for Seaweed Investment in Europe', available at www.seaweedeurope.com.

133 Central Intelligence Agency (CIA), *The World Factbook – Field listing – Coastline*, available at user.iiasa.ac.at/~marek/fbook/04/fields/2060.html.

134 Kraan, S. and Guiry, M.D., 'The Seaweed Resources of Ireland', *Seaweed Resources of the World* (2006).

135 Kim, J.K., Stekoll, M. and Yarish, C., 'Opportunities, challenges and future directions of open water seaweed aquaculture in the United States', *Phycologia*, 58 (2019).

136 MARINER (Macroalgae Research Inspiring Novel Energy Resources) ARPA-E, arpa-e.energy.gov/technologies/programs/mariner.

137 US Bureau of Labor Statistics. *Alaska Economy at a Glance* (2021), available at www.bls.gov.

138 Alaska Fisheries Development Foundation. *Alaska Mariculture Initiative*. Concept Paper (2014).

139 FAO, PAHO, WFP, UNICEF and IFAD, 'Regional Overview of Food Security and Nutrition in Latin America and the Caribbean 2020 – Food security and nutrition for lagged territories – In brief. Santiago' (2021).

140 Hanif, N., et al., 'L'exploitation des algues rouges *Gelidium* dans la région d'El-Jadida : aspects socio-économiques et perspectives' (2014).

141 www.agrifutures.com.au/wp-content/uploads/2020/09/20-072.pdf.

142 SeaMark – Ocean Rainforest, www.oceanrainforest.com/seamark.

143 World Economic Forum, *Why are most of the world's hungry people farmers?* (2015).

144 The National WWII Museum, article: There Are No Civilians in Japan (August 2020).

145 Neushul, P., Wang, Z., 'Between the Devil and the Deep Sea: C.K. Tseng, Mariculture and the Politics of Science in Modern China', *Isis*, Vol. 91, No. 1 (2000).

146 Froehlich, H.E., Afflerbach, J.C., Frazier, M., Halpern, B.S., 'Blue Growth Potential to mitigate climate change through seaweed offsetting', *Current Biology* (2019).

147 Duarte, C.M., Wu, J., Xiao, X., Bruhn, A., Krause-Jensen, D., 'Can Seaweed Farming Play a Role in Climate Change Mitigation and Adaptation?', *Frontiers in Marine Science*, 4 (2017).

148 www.worldwildlife.org/press-releases/wwf-receives-100-million-for-nature-based-climate-solutions-from-the-bezos-earth-fund.

149 Cury, P. and Pauly, D., *Mange tes méduses !*, Odile Jacob (2013).

150 Available at kelp.blue/.

151 www.acadianseaplants.com/land-based-seaweed-cultivation/.

152 Lavaut, E., Guillemin, M.-L., Colin, S., Faure, A., Coudret, J., Destombe, C. and Valero, M., 'Pollinators of the sea: A discovery of animal-mediated fertilization in seaweed', *Science* 377 (2022).

153 Schmidt, R., Saha, M., 'Infochemicals in terrestrial plants and seaweed holobionts: current and future trends', *New Phytologist* (2020).

154 Potin, P., 'Communiquer avec la chimie', *Pour la science*, No. 73 (November 1999).

155 Dechow, C.D., Liu, W.S., Specht, L.W. and Blackburn, H., 'Reconstitution and modernization of lost Holstein male lineages using samples from a gene bank', *Journal of Dairy Science*, Vol. 103, Issue 5 (2020).

156 Christopher Columbus, 'Journal of the First Voyage of Columbus', in *Journal of Christopher Columbus (during his first voyage, 149293), and Documents Relating to the Voyages of John Cabot and Gaspar Corte Real*, edited and translated by Markham, C.R. (London: Hakluyt Society, 1893), pp. 15-193.

157 *The Periplus of Hanno A Voyage Of Discovery Down The West African Coast, By A Carthaginian Admiral Of The Fifth Century B.C.*, translated from the Greek by Wilfred, H. and Schoff, A.M., available at archive.org/stream/cu31924031441847/cu31924031441847_djvu.txt.

158 *Festus Avienus raconte le voyage du carthagenois Hamilcon dans ses écrits*, available at issuu.com/scduag/docs/adg18129-1/15 – translator's own version from the French.

159 Jornandes, *Histoire des Goths (I-XXI) texte bilingue* (remacle.org) – translator's own version from the French.

160 etc.usf.edu/lit2go/83/twenty-thousand-leagues-under-the-sea/1442/part-2-chapter-11-the-sargasso-sea/.

161 Each of these species includes a wide variety of algal forms. *Sargassum natans* includes nine different forms whose genetic differences have yet to be clarified.

162 Smetacek, V. and Zingone, A., 'Green and golden seaweed tides on the rise', *Nature*, 504, pp. 84–88 (2013).

163 *Lobatus gigas* (queen conch), a marine mollusc found in equatorial and tropical areas. Heavily overfished in the West Indies and Florida because of the decorative use of its shell. Now classified as endangered.

164 United Nations, *The Second World Ocean Assessment*, Vol. I, p. 37 (2021).

165 De Morais Garofalo, L., *Les algues, des plantes marines peu utilisées au statut d'une 'algarchitecture'*, Master thesis, ENSA Paris-Malaquais (2022).

166 Desrochers, A., Cox, S-A., Oxenford, H.A. and van Tussenbroek, B., 'Pelagic sargassum – A guide to current and potential uses in the Caribbean', FAO Fisheries and Aquaculture Technical Paper No. 686. Rome, FAO (2022).

167 www.frontiersin.org/articles/10.3389/fmars.2021.618950/full#:~:text=%E2%80%9CGreen%20tide%E2%80%9D%20is%20a%20phenomenon,coastal%20zones%20(Table%201).

168 *INSEE Analysis* No. 32. 'La Bretagne : première région française pour la production et la transformation de viande' (January 2016), available at www.insee.fr/fr/statistiques/1908482.

169 Léraud, I. and Van Hove, P., *Algues vertes, l'histoire interdite*, (comic), Delcourt (2019).

170 Global Agricultural Information Network (GAIN), Report Name: Dutch Government Announces Programs to Curb Nitrogen Emission (May 2020).

171 He, Q., Bertness, M., Bruno, J., et al., 'Economic development and coastal ecosystem change in China', *Scientific Reports*, 4 (2014).

172 Erik Orsenna, *Petit Précis de mondialisation II : L'avenir de l'eau*, Fayard (2008).

173 Zhang, J., Wu, W., Ren, J., Lin, F., 'A model for the growth of mariculture kelp Saccharina japonica in Sanggou Bay, China', *Aquaculture Environment Interactions* (2016).

174 Xiao, X., Agusti, S., Lin, F., et al., 'Nutrient removal from Chinese coastal waters by large-scale seaweed aquaculture', *Scientific Reports*, 7, 46613 (2017).

175 Lomartire, S., Pacheco, D., Araujo, G., Marques, J., Pereira, L., Gonçalves, A., 'Wastewater Utilization as Growth Medium for Seaweed, Microalgae and Cyanobacteria, Defined as Potential Source of Human and Animal Services' (2021).

176 Lawton, R.J., Glasson, C.R.K., Novis, P.M. et al., 'Productivity and municipal wastewater nutrient bioremediation performance of new filamentous green macroalgal cultivars', *J. Appl. Phycol.*, 33 (2021).

177 Australian Seaweed Institute, *Seaweed Biofilters* (2021), available at www.australianseaweedinstitute.com.au/seaweed-biofilters.

178 Orland, C., Queiros, A., Spicer, J., Mcneill, C., Higgins, S., Goldworthy, S., Zananiri, T., Archer, L. and Widdicombe, S., 'Application of computer-aided tomography techniques to visualize kelp holdfast structure reveals the importance of habitat complexity for supporting marine biodiversity', *Journal of Experimental Marine Biology and Ecology* (2016).

179 Ifremer, *Production primaire du phytoplancton : vers une diminution globale ?* (2015).

180 Theuerkauf, S., Barrett, L., Alleway, H., Costa-Pierce, B., St. Gelais, A., Jones, R., 'Habitat value of bivalve shellfish and seaweed aquaculture for fish and invertebrates: Pathways, synthesis and next steps', *Reviews in Aquaculture*, 14 (2021).

181 Küpper, F.C., Carpenter, L.J., McFiggans, G.B., Palmer, C.J., Waite, T.J., Boneberg, E.M., Woitsch, S., Weiller, M.,

Abela, R., Grolimun, D., Potin, P., Butler, A., Luther III, GW., Kroneck, P.M.H., MeyerKlaucke, W., Feiters, M.C., 'Iodide accumulation provides kelp with an inorganic anti-oxidant impacting atmospheric chemistry', *Proceedings of the National Academy of Sciences of the USA* (2008).

182 Schmidt, G.A., Ruedy, R.A., Miller, R.L. and Lacis, A.A., 'Attribution of the present-day total greenhouse effect', *J. Geophys. Res.*, 115, D20106, (2010).

183 www.theguardian.com/tv-and-radio/2019/jan/21/david-attenborough-tells-davos-the-garden-of-eden-is-no-more.

Vincent Doumeizel is Senior Adviser on the oceans to the United Nations Global Compact as well as director of the Food Programme at the Lloyd's Register Foundation. A self-described optimist and global citizen, Vincent has in recent years devoted himself to promoting a food revolution and environmental solutions based on sea resources, especially seaweed.

Vincent leads the charitable objectives of the Foundation through the funding of innovative projects to drive safety in the food supply chain. Partnering with the UN, FAO, World Bank, WWF, universities, NGOs and large brands, he released the "Seaweed Manifesto" and now co-leads the Global Seaweed Coalition with the objective to scale up the seaweed industry safely in order to address some of the world's most important challenges, such as hunger, global warming, pollution and poverty.

Charlotte Coombe is an award-winning British translator working from French and Spanish into English since 2008 across a variety of genres including literary fiction and non-fiction. She has translated more than a dozen books by authors including Marvel Moreno, Margarita García Robayo, Ricardo Romero, Eduardo Berti and Abnousse Shalmani. Her translations of short stories and poetry have appeared in literary journals such as *The Southern Review*, *Modern Poetry in Translation*, *World Literature Today* and *Words Without Borders*. She is also the co-founder of Translators Aloud, a YouTube project that shines a light on translators reading from their work.